Guide to Fortran 2008 Programming

Walter S. Brainerd

Guide to Fortran 2008 Programming

Walter S. Brainerd
The Fortran Company
Tucson, Arizona, USA

ISBN 978-1-4471-6758-7 ISBN 978-1-4471-6759-4 (eBook)
DOI 10.1007/978-1-4471-6759-4

Library of Congress Control Number: 2015949998

Springer London Heidelberg New York Dordrecht

Printed on acid-free paper

Springer-Verlag London Ltd. is part of Springer Science+Business Media (www.springer.com)

Preface

Fortran has been the premier language for scientific computing since its introduction in 1957. Fortran originally was designed to allow programmers to evaluate formulas—FORmula TRANslation—easily on large computers. Fortran compilers are now available on all sizes of machines, from small desktop computers to huge multiprocessors.

The *Guide to Fortran 2008 Programming* is an informal, tutorial introduction to the most important features of Fortran 2008 (also known as Fortran 08), the latest standard version of Fortran. Fortran has many modern features that will assist the programmer in writing efficient, portable, and maintainable programs that are useful for everything from "hard science" to text processing.

Target Audience

This book is intended for anyone who wants to learn Fortran 08, including those familiar with programming language concepts but unfamiliar with Fortran. Experienced Fortran 90/95 programmers will be able to use this volume to learn more about the important features of F90/95, such as modules and arrays, and to assimilate quickly those features in Fortran 03 and Fortran 08 that are not in Fortran 90/95.

This guide is not a complete reference work for the entire Fortran language; it covers the basic features needed to be a good Fortran programmer and an introduction to the important features of Fortran 66, 77. 90, 95, 03, and 08. Many older error-prone features have been omitted and some of the more esoteric features that are new to Fortran also are not discussed. To understand some of the features used in old Fortran programs, other sources should be consulted after learning the best basic collection of features for writing new codes or enhancing old ones.

Guide to Fortran 2008 Programming is organized so that it may be read from beginning to end, but also particular topics may be studied by reading some chapters before previous ones are mastered. To a reasonable extent, all the material about one topic is presented together, making the book suitable as a reference work, as well as a tutorial.

Examples and Case Studies

Most of the important features of the Fortran programming language are covered with examples, beginning with the simplest constructs. The book concentrates to some extent on the newer features of the Fortran programming language, because they often provide the best facilities to accomplish a particular programming task. Both the style of the many example programs and the selection of topics discussed in detail guide the reader toward acquiring programming skills to produce Fortran programs that are readable, maintainable, and efficient.

Case studies are used to illustrate the practical use of features of Fortran 08 and to show how complete programs are put together. There are also simple problems to enable the reader to exercise knowledge of the topics learned.

Style of the Programming Examples

To illustrate the author's opinion of good Fortran programming style, the programming examples do not illustrate the wide variety of styles that might be used. There are certainly many other good programming styles, but it is important to use a style consistently within a programming project. The style notes also reflect opinions of the author and reflect one possible good style to use.

Most of the program examples have been run on either the free GCC compiler `gfortran` or the Intel Fortran compiler `ifort`.

Organization of the Content

An unusual feature of the book is that the first chapter contains a complete discussion of all the basic features needed to write complete Fortran programs: the form of Fortran programs, data types, simple expressions and assignment, and simple input and output. Subsequent chapters contain detailed discussions of control constructs, modules, procedures, arrays, character strings, data structures and derived types, pointer variables, input/output, object-oriented programming, and coarrays.

Module-oriented programming is a very important part of Fortran programming and the topic of modules is introduced early to provide the framework for organizing data and procedures for a Fortran program.

From the beginning, Fortran has had extensive facilities for input and output; however, this is a topic that is not explored fully in many books because it is a little more difficult than other features and perhaps just not as interesting as some features. The use of these facilities is very important in production programs, so this book contains, in Chapter 11, an extensive discussion of the excellent input/output facilities in Fortran.

Appendix A lists the many intrinsic procedures. Appendix B provides a brief informal syntax specification for the language.

There still will be occasions when more details about the language must be learned. In these cases it will be necessary to consult the official standard, published by the International Standards Organization or the reference work *The Fortran 2003 Handbook*, by Adams, Brainerd, Hendrickson, Maine, Martin, and Smith, Springer, 2009.

For more information about Fortran, go to `http://www.fortran.com`.

Many suggestions made by Brian Smith improved the book significantly.

Bill Long of Cray Inc. ran some programs using the Cray compiler that used Fortran 2008 features not implemented on any compiler available to the author.

Tucson, Arizona, USA Walter S. Brainerd

Contents

Introduction to Programming in Fortran 1

The best way to learn a programming language is to start reading and writing programs immediately. If a computer is available, we encourage you to write and *run* programs modeled on the simple programs in this chapter. In addition to this book, you will need a short set of directions to show you how to enter and run a program at your local installation.

This chapter covers the basic features of Fortran needed to perform simple calculations and print their results. Two case studies illustrate these features and provide some insight into the debugging process.

1.1 Programs that Calculate and Print

Since computers are very good at arithmetic and Fortran is designed to be very good at expressing numerical computations, one reasonable thing to learn first about Fortran is how to tell a computer to do the sort of arithmetic that otherwise might be done by hand or with the aid of a hand calculator. This section describes how to write programs to calculate and to print the answer.

Simple Calculations

The first example is a program that prints the result of an addition.

```
program calculation_1
   print *, 84 + 13
end program calculation_1
```

The program `calculation_1` tells the computer to add the numbers 84 and 13 and then to print the computed sum, 97. When the computer is told to run `calculation_1`, it does precisely that: It adds the two numbers and prints their sum. The execution output will look something like this.

```
97
```

Editing, Compiling, and Running a Program

Use your favorite editor on your computer system to edit a file with suffix specified by the compiler you are using, for example, `.f90`. How this is done varies from one system to another and it is assumed that you can do this. For our example, the file might be named `calculation_1.f90`. It is a good scheme to name the file the same as the program, but with the `.f90` suffix.

To see the contents of the file at any time, you can use the editor again or type a command at the prompt, which might be `more`, `less`, `type`, `cat`, or something else, depending on your system. For example, on a Linux system that uses $ as the prompt:

```
$ less calculation_1.f90

program calculation_1
   print *, 84 + 13
end program calculation_1
```

Compiling the program means invoking a piece of software (compiler) that translates the Fortran statements to computer instructions. This is done with a command similar to the following:

```
$ fortran calculation_1.f90
```

If the compilation is successful, an executable program called `a.out` or `a.exe` will be found in your directory or folder; you may confirm this by listing its contents (`ls` or `dir`, for example).

The program may now be run by typing the command a, `a.exe`, or `./a.out`.

```
$ ./a.out

  97
```

There are more sophisticated ways to edit and run a program, such as using a graphical interface supplied with many compilers.

Default Print Format

The asterisk following the keyword `print` tells the computer that the programmer will not be specifying the exact **format** or layout for the printed answer. Therefore, the Fortran system will use a **default format**, also called a list-directed format (11.8), designed to be satisfactory in most cases. The Fortran programming language allows some freedom in the design of default formats, so your output may differ slightly from the sample execution shown above.

Printing Messages

If you want the computer to print the exact typographic characters that you specify, enclose them in quotation marks (double quotes), as illustrated by the program `quotes`. The quotes are not printed in the output.

```
program quotes
   print *, "84 + 13"
end program quotes
```

```
 84 + 13
```

In a Fortran program, a sequence of typographic characters enclosed in quotes is a **character string.** A character string may contain alphabetic characters as well as numeric characters and may contain other special characters such as punctuation marks and arithmetic symbols.

Printing both exact literal characters and a computed numeric value produces the following easy-to-read output.

```
program calculation_1_v2
   print *, "84 + 13 =", 84 + 13
end program calculation_1_v2
```

```
 84 + 13 = 97
```

In the program `calculation_1_v2` (calculation 1 version 2), there are two items in the list in the `print` statement, a character constant `"84 + 13 ="` to be printed exactly as written (but without the delimiting quotation marks) and an arithmetic expression whose value is first calculated and then printed. Although the two items may look similar, they are treated quite differently. Enclosing the character string in quotes means that it is to be transcribed *character for character,* including the three blank characters (spaces, in ordinary typing), while the same expression written without quotes is to be evaluated so that the sum can be printed. Commas are used to separate the items in the list of a `print` statement.

The *program* Statement

Each Fortran program begins with a `program` statement and ends with an `end program` statement. The **`program` statement** consists of the keyword `program` followed by a **program name** of the programmer's choice. A name must start with a letter and consist of at most 63 letters, digits, and underscores; the letters may be uppercase or lowercase. Other names in Fortran also follow this rule.

The *end program* Statement

The **`end program` statement** begins with the keywords `end program`. It must be followed by the name of the program. Every Fortran program must have an `end program` statement as its last statement.

Exercises

1. Write and run a program that prints your name.

2. Write and run a program that computes the sum of the integers 1 through 9, preceded by a short message explaining what the output is.

3. What computer output might be expected when the following program is run?

```
program simple
   print *, 1, "and", 1, "equals", 1 + 1
end program simple
```

1.2 Intrinsic Data Types

The intrinsic (i.e., built-in) **data types** in Fortran are integer, real, complex, logical, and character. Each data type has a set of values that may be represented in that type and operations that can be performed on those values. We already have seen examples of the use of two of these data types. "84 + 13" (including the quotation marks) is a character string constant, and 84 + 13 is an expression whose value is of type integer, involving two integer operands, 84 and 13, and the arithmetic operator +. The following subsections discuss each of the five intrinsic types and the way that constants of those types are written in Fortran.

Integer Type

The **integer type** is used to represent values that are whole numbers. In Fortran, integer constants are written much like they are written in ordinary usage. An **integer constant** is a string containing only the digits 0–9, possibly followed by an underscore (_) and a named integer constant, which designates the kind parameter as described later in this section. The following are examples of integer constants.

```
23   0   1234567   42_short   42_long
```

Real Type

There are two forms of a **real constant** in Fortran. The first is called **positional form** because the place value of each digit is determined by its position relative to the decimal point. The positional form of a real constant consists of an integer followed by a decimal point followed by a string of digits representing the fractional part of the value, possibly followed by an underscore and a kind parameter. Assuming that `double` and `quad` are names of integer constants that are permissible real kinds on the Fortran system being used, all the following are real constants written in positional form.

```
13.5            0.1234567     123.45678
00.30_double    3.0           0.1234567_quad
```

The **exponential form** of a real number consists of a real number written in positional form followed by the letter `e` and an optionally signed integer (without a kind parameter) and optionally followed by an underscore and kind parameter. The letter `e` is read as "times 10 to the power" and the integer following the `e` is a power of 10 to be multiplied by the number preceding the `e`. Exponential notation is useful for writing very large or very small numbers. For example, `23.4e5` represents 23.4 times 10 to the power 5, 23.4×10^5, or $23.4 \times 100{,}000 = 2{,}340{,}000$. The integer power may contain a minus or plus sign preceding it, as in the real constant `2.3e-5`, which is 2.3×10^{-5} or $2.3 \times 0.00001 = 0.000023$. Two more examples are `1.0e9_double`, which is one billion with kind parameter `double`, and `1.0e-3`, which is 1/1000.

Complex Type

The Fortran **complex type** is used to represent the mathematical complex numbers, which consist of two real numbers and often are written as $a + bi$. The first real number is called the **real part** and the second is called the **imaginary part** of the complex number. In Fortran, a **complex constant** is written as two (possibly signed) real numbers, separated by a comma and enclosed in parentheses. If one of the parts has a kind parameter, the other part should have the same kind parameter; the complex constant then is that kind. Examples of complex constants are

```
(1.0, -1.0)
(-1.0, 3.1e-27)
(3.14_double, -7.0_double)
```

In the last example, `double` must be an integer parameter whose value is a kind available on the system being used.

Arithmetic Operators

The operators that may be used to combine two numeric values (integer, real, or complex) include +, -, *, /, and **. Except for **, these symbols have their usual mathematical meaning indicating addition, subtraction, multiplication, and division. The two asterisks indicate exponentiation; that is, the value of `2**4` is 16, computed as 2 raised to the power 4 or 2^4 in mathematical notation. The symbols + and - may be used as unary operators to indicate the identity and negation operations, respectively.

Integer division always produces an integer result obtained by chopping off any fractional part of the mathematical result. For example, since the mathematical result of 23/2 is 11.5, the value of the Fortran arithmetic expression

```
23.0 / 2.0
```

is 11.5, but the value of the expression

```
23 / 2
```

which is the quotient of two integer constants, is 11. Similarly, the value of both the expressions

```
-23 / 2        23 / (-2)
```

is −11.

Relational Operators

Numeric (and character) values may be compared with **relational operators**. The form of each relational operator is given in Table 1-1. Complex values may be compared

Table 1-1 The relational operators

Fortran form	Meaning
<	Less than
<=	Less than or equal to
==	Equal to
/=	Not equal to
>=	Greater than or equal to
>	Greater than

only with the relational operators == (equal) and /= (not equal). However, due to roundoff error, in most cases it is not appropriate to compare either real or complex values using either the == or the /= operator. In such cases, it is better to test for approximate equality instead. For example, it is possible to check that x is approximately equal to y with the expression

```
abs(x - y) < 1.0e-5
```

where `abs(x - y)` is the absolute value of the difference between x and y. The result of a relational operator is type logical.

Mixed-Mode Expressions

Mathematically, the integers are a subset of the real numbers and the real numbers are a subset of the complex numbers. Thus, it makes sense to combine two numeric values, creating a **mixed-mode expression**, even if they are not the same type. The two operands of a numeric operator do not have to be the same data type; when they are different, one is converted to the type of the other prior to executing the operation. If one is type integer and the other is type real, the integer is converted to a real value; if one is type integer and the other is type complex, the integer is converted to a complex value; if one is type real and the other is type complex, the real is converted to a complex value. As an example, the value of the expression

```
23.0 / 2
```

is 11.5, because the integer 2 is converted to a real value and then a division of two real values is performed. If the two operands have different kind parameters, usually the number whose kind parameter specifies lesser precision is converted to the kind with greater precision before the operation is performed.

Logical Type

The **logical type** is used to represent the two truth values true and false. A **logical constant** is either `.true.` or `.false.`, possibly followed by an underscore and a kind parameter.

The operators that may be used to combine logical values are `.not.`, `.and.`, `.or.`, `.eqv.`, and `.neqv.` They are all binary operators except the unary operator `.not.` The value resulting from the application of each logical operator is given in Table 1-2. To give one simple example, the value of

```
.false. .eqv. .false.
```

is true.

Table 1-2 Values of the logical operators

x_1	x_2	.not. x_1	x_1 .or. x_2	x_1 .and. x_2	x_1 .eqv. x_2	x_1 .neqv. x_2
True	True	False	True	True	True	False
True	False	False	True	False	False	True
False	True	True	True	False	False	True
False	False	True	False	False	True	False

Character Type

The **character type** is used to represent strings of characters. The form of a **character constant** is a sequence of any characters representable in the computer delimited by quotation marks. If a quotation mark is to occur in the character string, it is represented by two quotation marks with no intervening characters. If the character constant is not default kind, the kind precedes the constant (see the third example below).

```
"Joan"
"John Q. Public"
iso_10646_"Don't tread on me."
"He said, ""Don't tread on me."""
```

There is only one character operator that produces a character result: **concatenation**. The symbol used is // and the result of the binary operator is a string of characters consisting of those in the first string followed by those in the second string. For example, the value of `"John Q." // "Public"` is the string `John Q.Public`. Note that there is no blank after the period, although there could have been; the value of `"John Q. " // "Public"` is the string `John Q. Public`.

Relational operators may be used to compare character values, which is done using the character collating sequence for default character kinds.

Parameters/Named Constants

A **parameter** is a **named constant**. Each parameter must be declared in a **type statement.** Type statements appear between the `program` statement and the beginning of the executable part of the program. Type statements also are used to give names to variables (1.3) and indicate their data type. Each parameter declaration consists of a keyword specifying a type, followed by a comma and the keyword `parameter`, followed by two colons. To the right of the double colon is a list of names, each followed by an assignment and the expression giving the parameter value. The initialization assignments are separated by commas. For example,

```
real, parameter :: pi = 3.14159, e = 2.71828
integer, parameter :: number_of_states = 50
```

declare `pi` and `e` to be real parameters and `number_of_states` to be an integer parameter with the value 50.

The value of a parameter is fixed by its declaration and cannot change during the execution of a program.

A parameter name may be used in every place in a program where the corresponding constant may be used; this is why it is also called a named constant. In addition, a parameter may be used in some places where a variable may not be used. Examples are indicating the size of a static array and the values selected by a `case` statement.

```
program parameter_example
   integer, parameter :: &
      number_of_states = 50, &
      number_of_senators_per_state = 2, &
      number_of_senators = &
         number_of_states * number_of_senators_per_state

   print *, &
      "There are", number_of_states, &
      "states in the United States of America."
   print *, &
      "From this, we can calculate that there are"
   print *, number_of_senators, &
      "senators in the United States senate."

end program parameter_example
```

The ampersand (&) indicates that a statement is continued on the next line (1.4).

Style note: It is good programming practice to declare quantities to be parameters whenever possible. Assigning a constant value to a parameter tells the reader of the program that the value corresponding to that name will never change when the program is running. It also allows the computer to provide a diagnostic message if the programmer inadvertently tries to change its value.

Since parameters are named constants, use of a parameter name instead of the corresponding constant makes a program more readable. It is easy to forget what role an unnamed constant plays in a program.

Another important reason for using a parameter declaration is that the program can be modified very easily if the particular value represented by the parameter name needs to be changed. The programmer can then be sure that the constant will be correct whenever it is used throughout the program. For example, if Puerto Rico becomes the 51st state, the program `parameter_example` can be updated easily.

A rather different reason for using a parameter is that its value is known by the compiler and, therefore, can be used to indicate such things as the size of an array (4.1) or the kind of a real variable.

Enumerators

Using an enumerator is a way to declare some related parameters. The required `BIND(C)` attribute (8.10), gives them the same type and kind as a C enumerator, but it is not necessary to call a C program when using an enumerator.

```
enum, bind(c)
   enumerator :: error_unit
   enumerator :: std_input = 5, std_output
end enum
```

Unless otherwise indicated, the first parameter of the enumerator is set to 0, so `error_unit` is 0, `std_input` is 5, and `std_output` is 6.

Rules for Names

`number_of_states` and `number_of_senators` are **names** of parameters used in the program `parameter_example`. The following are the rules for names of parameters as well as all other names in a Fortran program:

1. The first character of the name must be a letter.

2. The remaining characters may be any mixture of letters, digits, or underscore characters (_).

3. There may be at most 63 characters in a name.

   Style note: Names may contain both uppercase and lowercase letters, but a program should not contain two names that differ only in the case of some of their letters. For example, a variable could be `Number_of_States`, but wherever it is used in a program, it should have the "`N`" and "`S`" capitalized. The name `number_of_states` is the same variable, but looks to the reader like it might be a different variable.

These rules allow ordinary names like Lisa, Pamela, and Julie to be used as names. They also allow ordinary English words such as total and area and more technical-looking names such as X3J3 and WG5 to be used as names. The underscore allows longer names to be more readable, as in distance_to_the_moon, vowel_count, and number_of_vowels_in_the_text.

All names in Fortran, including names of programs, follow these rules.

Most names in this book are all lowercase, simply because they are a little easier to type.

Kind Parameters

Kind parameters provide a way to parameterize the selection of different possible machine representations for each of the intrinsic data types. If the programmer is careful, this provides a mechanism for making selection of numeric precision and range portable.

Each intrinsic data type has a parameter, called its **kind parameter**, associated with it. A kind parameter is intended to designate a machine representation for a particular data type. As an example, an implementation might have three real kinds, informally known as single, double, and quadruple precision.

The kind parameter is an integer. These numbers are processor dependent, so that kind parameters 1, 2, and 3 might be single, double, and quadruple precision; or on a different system, kind parameters 4, 8, and 16 could be used for the same things. There are at least two real and complex kinds and at least one kind for the integer, logical, and character data types. There must be an integer kind capable of representing all 18-digit integers. Note that the value of the kind parameter is not usually the number of decimal digits of precision or range; on many systems, it is the number of bytes used to represent the value.

It is possible to determine which kind parameters are available for each type on your system by using the parameters real_kinds, integer_kinds, logical_kinds, and character_kinds in the intrinsic module iso_fortran_env.

```
program kinds

   use iso_fortran_env
   implicit none
   print *, real_kinds

end program kinds
```

Kind parameters are optional in all cases, so it is possible to always use the default kind if that is sufficient for your application.

The intrinsic functions selected_int_kind and selected_real_kind may be used to select an appropriate kind for a variable or a named constant. These functions provide the means for making a program portable in cases where values need to be computed with a certain specified precision that may use single precision on one machine, but require double precision on another machine.

When a kind parameter is used in a program, it must be a named integer constant (parameter). In integer, real, and logical constants, it follows an underscore character (_) at the end.

```
12345_short
1.345_very_precise
.true._enough
```

The two operands of a numeric operation may have different kind parameter values. In this case, if the two operands have the same type or one is real and one complex, the result has the kind parameter of the operand with the greater precision. For example, if kind `long` has greater precision than kind `short`, the value of

```
1.0_short + 3.0_long
```

is 4.0 with kind parameter `long`. If one operand is type integer and the other is real or complex, the kind parameter of the result is that of the real or complex operand.

Exercises

1. Convert the following type real numbers from positional notation to exponential notation.

```
48.2613      0.00241_ok      38499.0
0.2717       55.0            7.000001_quad
```

2. Convert the following type real numbers from exponential notation to positional notation.

```
9.503e2      4.1679e+10_double     2.881e-5
-4.421e2     -5.81e-2_nice         7.000001e0
```

3. Write a program that prints the sum $0.1 + 0.2 + 0.3 + \cdots + 0.9$.

4. Determine the kind number of one real kind that has precision greater than that of the default real kind on your computer system.

5. Print the value of `selected_int_kind` and `selected_real_kind` for about a dozen different argument values to see which kind values are available on the computer you are using. Check your results with your compiler manual.

6. Write a program that prints the sum of the complex numbers $(0.1+0.1i) + (0.2+0.2i) + (0.3+0.3i) + (0.4+0.4i)$.

7. Write a program that prints the logical value of each of the following expressions:

```
2 > 3
2 < 3
0.1 + 0.1 == 0.2
0.5 + 0.5 /= 1.0
```

8. Write a program that computes and prints the concatenation of all your names (i.e., first, middle, and last).

1.3 Variables and Input

One benefit of writing a computer program for doing a calculation rather than obtaining the answer using pencil and paper or a hand calculator is that when the same sort of problem arises again, the program already written can be reused. The use of variables gives the programs in this section the flexibility needed for such reuse. The programs in 1.1 direct the computer to perform the indicated arithmetic operations on numeric constants appearing in the `print` statements. The sample program `add_2` finds the sum of any two integers supplied as input. The numbers to be added do not appear in the program itself. Instead, two integer variables `x` and `y` are reserved to hold the two values supplied as input. Because Fortran statements can operate on variables as well as constants, their sum can be calculated and printed. The first sample run shows how this new program could be used to find the sum of the numbers 84 and 13, calculated by the program `calculation_1` in 1.1.

```
program add_2
   integer :: x, y
   read *, x
   print *, "Input data  x:", x
   read *, y
   print *, "Input data  y:", y
   print *, "x + y =", x + y
end program add_2
```

```
 Input data  x: 84
 Input data  y: 13
 x + y = 97
```

After declaring that the variables `x` and `y` will hold integer values, the program `add_2` tells the computer to read a number from an input device and call it `x`, then to read another number and call it `y`, and finally to print the value of `x + y`, identified as such. Two additional `print` statements that echo the values of the input data complete the program `add_2`. During the execution of this program, the two numbers that are the values for `x` and `y` must be supplied to the computer, or the computer cannot complete the run.

Declaration of Variables

The value of a parameter is fixed by its declaration and cannot change during execution of a program. On the other hand, if the keyword `parameter` is omitted, the objects being declared become variables and their values can be changed at any time. Thus,

```
integer :: count
```

declares count to be an integer variable. The value of a variable declared in this way may be changed during execution of the program.

The program add_2 uses the type declaration

```
integer :: x, y
```

that declares the type of the variables x and y.

Variable names are subject to the same rules as parameter names.

Corresponding to the integer, real, complex, logical, and character constants introduced in 1.2, there are integer, real, complex, logical, and character variables. For example, if the variables q, t, and k are to be real variables in a program and the variables n and b are to be integer variables, then the following lines contain the necessary declarations.

```
real :: q, t, k
integer :: n, b
```

Variables may have a particular hardware representation by putting kind= followed by a named constant in parentheses after the keyword representing the data type. For example, if more significant digits are needed than your system keeps in the default real type and the kind parameter for extra precision is 2, the variables dpq, x, and long may have extra precision by the following declarations.

```
integer, parameter :: more_precision = 2
real(kind=more_precision) :: dpq, x, long
```

A character variable may have a kind parameter, but it should have a **length**. The keyword character should be followed by parentheses enclosing len= and an integer value indicating the number of characters in the character string. If no length is given, it is assumed to be 1. If the variable name is to be a string of 20 characters, it may be declared as follows.

```
character(len=20) :: name
```

Instead of an integer value, the length can be * (meaning "assumed" or specified elsewhere) for a character parameter or a dummy argument (3.5) that is type character. It also may be :, which indicates that the length of the string is determined at run time; in this case, the string must also have the allocatable (5.1) attribute.

The type statement (6.3) may be used to declare things to be an intrinsic type.

```
type (real) :: x, y, z
type (integer(kind=long_int_kind)), parameter :: &
      large_integer = 99999999
```

The *implicit none* Statement

If a variable does not appear in a type declaration statement, it has a type that is determined by its first letter (variables with names beginning with I, J, K, L, M, or N are type integer; all others are type real). This is very error prone and can be avoided by putting the statement

```
implicit none
```

before the type declarations of every program.

Style note: Every program and module should contain an `implicit none` statement. This requires that every variable that is used in the program must be listed in a type declaration.

Supplying Input Data

The two input values 84 and 13 for the variables `x` and `y`, shown in the sample execution of the program `add_2`, did not appear in the computer by magic. They were typed in by the user, but not as part of the program file. Instead, an input file can be prepared, usually with the same editor used for preparing the program file. In this case the file contains the two lines

```
84
13
```

If, for example, the file is named `add_2.in`, the program can be executed on most computer systems using a command similar to the following:

```
$ ./a.out < add_2.in
```

If you want to put the output in a file called `add_2.out` instead of displaying it on your screen, the following command should work:

```
$ ./a.out < add_2.in > add_2.out
```

Echo of Input Data

When reading input data from a file, it is good programming practice for the user to provide an **echo of the input data** using `print` statements, so that the output contains a record of the values used in the computation. Each `read` statement in the program `add_2` is followed by an echo of the input data just read.

Style note: It is good programming practice to echo all input data read from an input file. However, it will be impractical to follow this rule in some cases, such as when there is a large amount of input data.

Rerunning a Program with Different Data

The program `add_2` contains echoes, whose importance is demonstrated when the program is rerun using different input data. The echoes of input data help identify which answer goes with which problem. Other important uses of input echoes will appear later. In showing another sample run of the program `add_2`, this time adding two different numbers, we do not repeat the program listing. The program does not change; only the input data change. This time, the data file `add_2.in` has the following two lines.

```
4
7
```

The execution output might look like

```
Input data  x: 4
Input data  y: 7
x + y = 11
```

The final `print` statement of `add_2` refers to the variables `x` and `y`. As the execution output for the two sample runs shows, what actually is printed is the value of the character string constant `"x + y = "` followed by the value of the expression `x + y` at the moment the `print` statement is executed.

The program `add_2_reals` is obtained from the program `add_2` simply by changing the keyword `integer` in the variable declaration to the keyword `real`, which causes the type of the variables `x` and `y` to be real. It can be used to add two quantities that are not necessarily whole numbers. This execution of the program also illustrates that the input data values may be negative. The input file `add_2_reals.in` for this sample execution contains two lines:

```
97.6
-12.9
```

The program file contains the following lines:

```
program add_2_reals
   implicit none
   real :: x, y
   read *, x
   print *, "Input data  x:", x
   read *, y
   print *, "Input data  y:", y
   print *, "x + y =", x + y
end program add_2_reals
```

and the output is as follows:

```
Input data  x:  97.5999985
Input data  y: -12.8999996
x + y =  84.6999969
```

Some Fortran systems habitually print real quantities in exponential format. On such a system, the sample execution will more closely resemble the following:

```
Input data  x:  0.975999985E+02
Input data  y: -0.128999996E+02
x + y =  0.846999969E+02
```

If you are worried about why the printed result is not exactly 84.7, see 1.7 about roundoff error.

Reading Several Values

The `read` statement may be used to obtain values for several variables at a time, as shown in the program `average`, that calculates the average of any four numbers. The four numbers to be averaged are supplied as data, rather than appearing as constants in the program. This permits the same program to be used to average different sets of four numbers.

```
program average
   implicit none
   real :: a, b, c, d
   read *, a, b, c, d
   print *, "Input data  a:", a
   print *, "            b:", b
   print *, "            c:", c
   print *, "            d:", d
   print *, "Average =", (a + b + c + d) / 4
end program average
```

The input data file in the sample execution has one line:

```
58.5 60.0 61.3 57.0
```

When we run the program `average` using this data file, the following output is produced.

```
 Input data  a:  58.5000000
             b:  60.0000000
             c:  61.2999992
             d:  57.0000000
 Average =  59.2000008
```

This program does a computation more complicated than any discussed so far, but the meaning of the program should be obvious.

As shown in the sample execution, the data are supplied to the variables in the order they are listed in the `read` statement. Note that the four variables in the `read` statement are separated by commas and that there is a comma between the asterisk and the first variable in the input list. Although it is not required, it is often desirable to put all input data for a `read` statement on one line in the input file, creating a correspondence between `read` statements and data lines. However, the input data file

```
58.5
60.0
61.3
57.0
```

would have produced the same execution output.

Execution of each **read** *statement normally reads data from a new line in the input file.* Thus, if four separate read statements were to be used to read the variables a, b, c, and d, the four input values must be on four separate data lines in the input file.

Default Input Format

The asterisk in the read statement indicates that the format of the input data is left to the one who prepares the input file, except that the individual values must be separated by at least one blank character or a comma.

Style note: Whenever possible, use the default input format. It makes preparation of data much easier and less prone to error.

Reading and Writing Character Strings

Since computers can process character data as well as numeric information, computer languages provide for the reading and printing of character strings. The somewhat facetious program who shows how this is done in Fortran.

```
program who
   implicit none
   character(len=20) :: whats_his_name

   print *, "Do I remember whatshisname?"
   read *, whats_his_name
   print *, "Of course, I remember ", whats_his_name
end program who
```

```
 Do I remember whatshisname?
 Of course, I remember Roger Kaputnik
```

When the default input format, indicated by the asterisk, is used to read a character string, you should enclose the string in quotes, the same as a character constant used within a program. Delimiting quotes do not appear in the output when using the default output format. The input file for the execution of the program who shown above consists of one line.

```
"Roger Kaputnik"
```

Input Data from a Terminal

We close this section with a program meters_to_inches designed to be run on a system in which input data is supplied for the read statements by typing the data at a computer terminal *during* the execution of the program. This is called **interactive input**. The only change we make to the program is to add a print statement prompting the user about what data to type and remove the statement that echoes the input data. This **input prompt** immediately precedes the read statement. Without this input prompt, when the computer pauses waiting for the user to type the value for meters requested in the read statement, it would appear as though the execution of the program meters_to_inches failed for some unexplained reason, or that it never started. The user would not know that the computer is waiting for input.

Style note: Precede an interactive input statement with an input prompt.

```
program meters_to_inches
! Converts length in meters to length in inches.
! The length in meters is typed
! when prompted during execution.

   implicit none
   real :: meters
   real, parameter :: inches_per_meter = 39.37

   print *, "Enter a length in meters."
   read *, meters
   print *, meters, "meters =", &
      meters * inches_per_meter, "inches."
end program meters_to_inches
```

```
 Enter a length in meters.
 2
   2.0000000 meters =  78.7399979 inches.
```

On most systems, the characters typed at the keyboard also appear on the screen.

Nonadvancing input/output allows the input to be typed on the same line as the prompt. There is an example of this in 11.3.

Exercises

1. Which of the following are valid names for Fortran variables?

```
name     address    phone_#     phoney     real
iou_     iou_2      4gotten     4_ever     _laurie
```

2. The program inches_to_feet is similar to the program meters_to_inches described in this section. What output is produced when inches_to_feet is run using 110 inches as the input value?

```
program inches_to_feet
   implicit none
   real :: inches
!  There are 12 inches per foot
   real, parameter :: inches_per_foot = 12.0

   read *, inches
   print *, inches, "inches =",  &
         inches / inches_per_foot, "feet."
end program inches_to_feet
```

3. In the program rhyme, both `jack` and `jill` are parameters. What does a computer print when this program is run?

```
program rhyme
   implicit none
   integer, parameter :: jack = 1, jill = 2
   print *, jack + jill, "went up the hill."
end program rhyme
```

4. Write a program that reads in a first name, a middle initial, and a last name as the values of three different character variables and prints out the full name.

1.4 The Form of a Fortran Program

A Fortran program consists of a sequence of statements; these statements are written on lines that may contain from 0 to 132 characters.

Continued Statements

Often a statement fits on one line, but a statement can be continued onto more lines if the last character of the line to be continued is an ampersand (&).

```
print *,  &
      "I hope this is the right answer."
```

A statement may not have more than 256 lines.

A statement should not be broken in the middle of a keyword, a name, or a constant. If it is necessary to break a long character string, use the concatenation operator as shown in the following example.

```
print *,  &
      "This is a line that contains a really, " // &
      "really, really, long character string."
```

The important fact is that, in the absence of a continuation symbol, the end of a line marks the end of a statement.

Each Fortran statement except the assignment statement (and the statement function statement, not discussed in this book) begins with a **keyword**, such as `print`, that identifies the kind of statement it is.

Significant Blank Characters

Blank characters are significant in a program. In general, they must not occur within things that normally would not be typed with blanks in English text, such as names and numbers. On the other hand, they must be used between two things that look like "words". An example is that, in the first line of a program, the keyword program and the name of the program must be separated by one or more blanks, as in the example

```
program add_2
```

Keywords and names such as print and number must contain no blank characters, except that keywords consisting of more than one English word may contain blanks between the words, as in the statement

```
end do
```

Two or more consecutive blanks are always equivalent to one blank unless they are in a character string.

On the other hand, there are places where blank characters are not significant, but can and should be used to improve the readability of the program. For example, many of the programs in this book have blanks surrounding operator symbols, such as + and -, and have a blank after each comma in an input/output list or procedure argument list. Even more importantly, they all use preceding blanks to produce indentation that shows the structure of the program and of its component parts.

Style note: Blank characters and blank lines should be used freely in a program to make it easier to read.

Comments

Any occurrence of the exclamation symbol (!) other than within a character string or a comment marks the beginning of a **comment**. The comment is terminated by the end of the line. All comments are ignored by the Fortran system and are used to provide the human reader information about the program.

Since comments are ignored, it is permissible to place a comment after the ampersand (&) continuation symbol without impairing the continuation.

```
real :: x,  &     ! measured value
        xbar      ! smoothed value
```

The Fortran Character Set

A Fortran statement is a sequence of characters. The characters of the Fortran character set are the uppercase letters A–Z, the lowercase letters a–z, the digits 0–9, the underscore _, and the special characters in Table 1-3. In addition, the character set contains the following characters, which have no special use in a Fortran program and may appear in a program only within a comment or character constant.

```
$   ~   ^   \   {   }   '   |   #   @
```

The currency symbol need not display or print as $ in all implementations; it might look like ¥ or £.

Table 1-3 The Fortran special characters

Character	Name of character	Character	Name of character
	Blank	;	Semicolon
=	Equals	!	Exclamation point
+	Plus	"	Quotation mark or quote
-	Minus	%	Percent
*	Asterisk	&	Ampersand
/	Slash	~	Tilde
(	Left parenthesis	>	Greater than
)	Right parenthesis	?	Question mark
[	Left bracket	'	Apostrophe
]	Right bracket	`	Grave accent
,	Comma	$	Currency symbol
.	Decimal point or period	#	Number sign
:	Colon	@	Commercial at

Exercise

1. What does the following program print? Its style is *not* recommended.

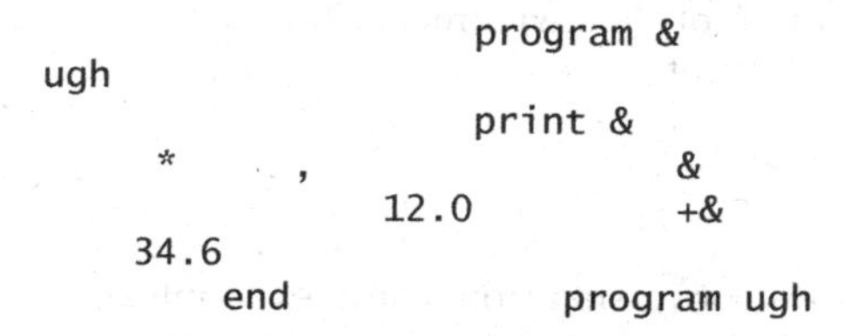

```
                         program &
ugh
                         print &
     *     ,                    &
                 12.0           +&
     34.6
        end                          program ugh
```

1.5 Some Intrinsic Functions

There are many **built-in** or **intrinsic functions** in Fortran, a few **built-in** or **intrinsic subroutines**, and a few **built-in** or **intrinsic modules**. To use the functions, simply type the name of the function followed by the arguments to the function enclosed in parentheses. For example, `abs(x)` produces the absolute value of `x` and `max(a,b,c)` yields the maximum of the values of `a`, `b`, and `c`.

Two of the more commonly used built-in subroutines are `date_and_time` and `random_number`. Appendix A contains a list of all the intrinsic procedures.

There are built-in modules for features that assist calling C programming language features (8.10) and handling IEEE arithmetic and exceptions (7).

Numeric Type Conversion Functions

There are built-in functions that convert any numeric value to each of the numeric types. These functions are named `int`, `real`, and `cmplx`. For example, the value of `int(4.7)` is the integer 4, the value of `real((2.7,-4.9))` is 2.7, the real part of the complex number 2.7–4.9*i*, and the value of `cmplx(2)` is 2.0+0.0*i*. These functions are essential in some situations, such as when it is necessary to convert an integer to a real to avoid an integer division or when the type of a procedure actual argument must match the type of a dummy argument. For example, if a variable `total` holds the sum of a bunch of integer test scores and it is necessary to divide by the integer variable `number_of_scores` to find the average, one or both must be converted to type real. Otherwise, the result will be an integer, which is probably not what is desired. The expression

```
real (total) / number_of_scores
```

will produce a real result with the fractional part of the average retained.

In other cases, explicit conversion is not required, but can improve the clarity of the program. For example, if `i` is an integer variable and `r` is a real variable, the assignment of the value of `r` to the variable `i` can be done with the statement

```
i = r
```

When this is done, any fractional part of the value of `r` is dropped, so that if `r` were 2.7, the value of `i` would be 2 after execution of the assignment. This can be made clearer to the reader of the program if the statement

```
i = int(r)
```

is used instead.

> *Style note:* In a context that requires conversion from complex to integer or real or requires conversion from real to integer, use the intrinsic type conversion functions even if they are not required.

The numeric type conversion functions also may be used to convert from one kind to another within the same data type or to specify the kind parameter of the result of conversion between data types. For example, `int(x, kind=short)` converts the real value `x` to an integer with kind parameter `short`. The kind should be given as an integer parameter.

The `logical` Function

The function named `logical` converts from one logical kind to another. For example, if `truth` is type logical and `packed` is an integer named constant, `logical(truth,` packed) is the value of truth represented as a logical with kind parameter packed and logical(truth) is the value of truth represented as a logical with the default kind parameter.

Mathematical Functions

There are several built-in functions that perform common mathematical computations. The following is a list of some of the most useful ones. Appendix A should be consulted for a complete list of the functions. Most of them do what would be expected, but the functions max and min are a little unusual in that they may be used with an arbitrary number of arguments. The mathematical functions are shown in Table 1-4.

Table 1-4 Mathematical intrinsic functions

abs	cos	min
acos	cosh	modulo
aimag	exp	sin
asin	floor	sinh
atan	log	sqrt
ceiling	log10	tan
conjg	max	tanh

Some of these functions will be used in the case studies at the end of this chapter. Other intrinsic functions, such as those for array processing and character processing, will be discussed in relevant chapters.

Kind Intrinsic Functions

The kind function returns the kind parameter value of its argument; the value depends on the integers used as kind parameters on the computer being used. For example, kind(x) is the kind parameter of the variable x; it might be 1 or 4, for example. kind(0) is the default integer kind; kind(0.0) is the default real kind; and kind(.false.) is the default logical kind.

There is an intrinsic function selected_real_kind that produces a kind value whose representation has at least a certain precision and range. For example, selected_real_kind(8, 70) will produce a kind (if there is one) that has at least 8 decimal digits of precision and allows a range of values between -10^{70} and -10^{-70}, values between $+10^{-70}$ and $+10^{70}$, and zero. This permits the programmer to select representations having required precision or range and give these processor-dependent kind values to named constants. The named constants can then be used to indicate the kind of a variable.

For the integer data type, there is an intrinsic function `selected_int_kind` with only one argument. For example, `selected_int_kind(5)` produces an integer representation allowing all integers between (but not necessarily including) -10^5 and $+10^5$.

Exercises

1. Write a program that prints the kind of each of the constants

   ```
   0
   0.0
   (0.0, 0.0)
   .false.
   "a"
   ```

   These are the default kinds.

2. Using the fact that `selected_real_kind` and `selected_int_kind` return a negative value when asked to produce a kind number for a precision or range not available on the system, determine all the possible kind numbers for reals and integers on your system.

1.6 Expressions and Assignment

A Fortran **expression** can be used to indicate many sorts of computations and manipulations of data values. So far we have seen simple examples of expressions as values to be printed using the `print` statement. We now discuss in more detail just what can appear in this list of things to be printed.

Primaries

The basic component of an expression is a **primary**. Primaries are combined with operations and grouped with parentheses to indicate how values are to be computed. A primary is a constant, variable, function reference (3.6), array element (4.1), array section (4.1), structure component (6.1), substring (5.1), array constructor (4.1), structure constructor (6.1), or an expression enclosed in parentheses. Note that this is a recursive definition because the definition of an expression involves expressions in parentheses. Examples of primaries are

```
5.7e43_double      ! constant
number_of_bananas ! variable
abs(x)             ! function value
(a + 3)            ! expression enclosed in parentheses
```

Primaries can be combined using the operators discussed in 1.2 as well as with user-defined operators discussed in 9.3 to form more complicated expressions. Any expression can be enclosed in parentheses to form another primary. Examples of more complicated expressions are

```
-a + d * e + b ** c
x // y // "abcde"
(a + b) /= c
log_1 .and. log_2 .eqv. .not. log_3
a + b == c * d
```

The Interpretation of Expressions

When more than one operation occurs in an expression, parentheses and the **precedence** of the operations determine the operands to which the operations are applied. Operations with the highest precedence are applied first to the operand or operands immediately adjacent to the operator. For example, since * has higher precedence than +, in the expression `a + b * c`, the multiplication is first applied to its operands `b` and `c`; then the result of this computation is used as an operand by adding it to the value of `a`. If the programmer intends to add `a` and `b` and multiply the result by `c`, parentheses must be used as in the expression `(a + b) * c`.

When two operators have the same precedence, they are applied left-to-right, except for exponentiation, which is applied right-to-left. Thus, the value of `9 - 4 - 3` is 5 – 3 = 2, but the value of `2 ** 3 ** 2` is $2^9 = 512$.

Table 1-5 shows the operations with the highest precedence at the top of the list and the ones with the lowest precedence at the bottom.

Table 1-5 Operator precedence

Operator	Precedence
User-defined unary operation	Highest
**	.
* or /	.
Unary + or -	.
Binary + or -	.
//	.
==, /=, <, <=, >, >=	.
`.not.`	.
`.and.`	.

Table 1-5 *(Continued)* Operator precedence

Operator	Precedence
`.or.`	.
`.eqv.` or `.neqv.`	.
User-defined binary operation	Lowest

The Evaluation of Expressions

Once it is determined by use of parentheses and precedence of operations which operations are to be performed on which operands, the computer may actually evaluate the expression by doing the computations in any order that is mathematically equivalent to the one indicated by the correct interpretation, *except that it must evaluate each subexpression within parentheses before combining it with any other value.* For example, the interpretation of the expression `a + b + c` indicates that `a` and `b` are to be added and the result added to `c`. Once this interpretation is made, it can be determined that a mathematically equivalent result will be obtained by first adding `b` and `c` and then adding this sum to `a`. Thus, the computer may do the computation either way.

The purpose of allowing the computer to rearrange expressions is to optimize execution speed. Thus a compiler will usually replace `x/2.0` with `0.5*x` because multiplication is faster than division on most computers. If execution speed is not important and you do not want to worry about these matters, just set the optimization level to 0 when compiling a program.

If the programmer writes the expression `(a + b) + c`, the computer must first do the computation as required by the parentheses. Note that the expression `(a + b) + (c + d)` can be done by first adding `c` and `d` but then the computer must add `a` and `b` and add that result to the first sum obtained. To evaluate this expression, the computer must not first add `b` and `c` or any other pair in which one operand is taken from `(a + b)` and the other is taken from `(c + d)`, because doing this would violate the integrity of parentheses.

Note that integer division is an oddity in that it does not satisfy the rules of arithmetic for ordinary division. For example, `(i / 2) * 2` is not equal to `i` if `i` is an odd integer. Thus, a computer may not make this substitution to optimize the evaluation of the expression.

Table 1-6 contains examples of expressions with allowable alternative forms that may be used by the computer in the evaluation of those expressions. `a`, `b`, and `c` represent arbitrary real or complex operands; `i` and `j` represent arbitrary integer operands;

`x`, `y`, and `z` represent arbitrary operands of any numeric type; and `l1`, `l2`, and `l3` represent arbitrary logical operands.

Table 1-6 Allowable alternative expressions

Expression	Allowable alternative form
`x + y`	`y + x`
`x * y`	`y * x`
`-x + y`	`y - x`
`x + y + z`	`x + (y + z)`
`x - y + z`	`(z - y) + x`
`x * a / z`	`x * (a / z)`
`x * y - x * z`	`x * (y - z)`
`a / b / c`	`a / (b * c)`
`a / 5.0`	`0.2 * a`
`i > j`	`j - i < 0`
`l1 .and. l2 .and. l3`	`l1 .and. (l2 .and. l3)`
`abs(i) > -1 .or. logical(l1)`	`.true.`

Table 1-7 contains examples of expressions with forbidden alternative forms that must not be used by a computer in the evaluation of those expressions.

Table 1-7 Nonallowable alternative expressions

Expression	Nonallowable alternative form
`i / 2`	`0.5 * i`
`x * i / j`	`x * (i / j)`
`(x + y) + z`	`x + (y + z)`
`i / j / a`	`i / (j * a)`
`(x * y) - (x * z)`	`x * (y - z)`
`x * (y - z)`	`x * y - x * z`

Assignment

The **assignment statement** is the most common way of giving a variable a value. An assignment statement consists of a variable, an equals sign (=), and an expression. The expression is evaluated and assigned to the variable. An example of an assignment statement is

```
x = a + 2 * sin(b)
```

Note for later that the variable on the left-hand side may be an array, an array element, an array section, a substring, or a structure component.

Complete agreement of the variable and expression type and kind is not always required. In some cases the data type or kind parameter of the expression may be converted in order to assign it to the variable.

If the variable on the left-hand side is any numeric type, the expression may be any numeric type and any kind. If the variable is type character of kind ASCII, ISO 10646, or default, the expression must be type character with one of those kinds; if the variable is type character with another kind, the expression must be character of the same kind. If the variable is type logical, the expression must be type logical but may be any kind. If the variable is a derived type (6.2), that is, a user-defined type, the expression must be the same derived type.

All of these rules apply to assignment as provided by the system (intrinsic assignment); it is possible to extend the meaning of assignment to other cases as described in 9.1.

Exercises

1. What computer output might be expected when the following program is run?

```
program calculation_2
   implicit none
   print *, (201 + 55) * 4 - 2 * 10
end program calculation_2
```

2. The program `calculation_3` uses a confusing sequence of arithmetic operations whose meaning would be clearer if written with parentheses. What computer output might be expected when it is run? Insert parentheses in the `print` statement in a way that does not change the value printed, but makes it easier to understand.

```
program calculation_3
   implicit none
   print *, 343 / 7 / 7 * 2
end program calculation_3
```

3. What computer output might be expected when `calculation_4` is run?

```
program calculation_4
   implicit none
   print *, 2 * (3 * (5 - 3))
end program calculation_4
```

4. What computer output might be expected when the program `power_of_2` is run?

```
program power_of_2
   implicit none
```

```
    print *, 2 * 2 * 2 * 2 * 2 * 2 * 2 * 2 * 2 * 2
end program power_of_2
```

5. Write an expression that rounds the value of the variable `x` to the nearest tenth.

6. When is `int(x/y)` equal to `x/y` for real values `x` and `y`?

7. If `x` and `y` are type integer and both are positive, the value of the intrinsic function `modulo(x, y)` is the remainder when `x` is divided by `y`. For example, `modulo(17, 5)` = 2. Rewrite the following expression using the built-in function `modulo`. Assume `n` is type integer with a positive value.

```
n - (n / 100) * 100
```

8. Write an expression using the built-in function `modulo` that has the value 1 when `n` is odd and 0 when `n` is even.

9. Write an expression using the built-in function `modulo` that is true if the value of the variable `n` is even and is false if `n` is odd.

10. Write a program to compute the quantity $e^{i\pi}$. The constant π can be computed by the formula 4 * `atan(1.0)` because $\tan(\pi/4) = 1$. The complex constant i can be written `(0.0, 1.0)`. The built-in function `exp(z)` is used for raising the mathematical constant e to a power. The sample output should look like

```
The value of e to the power i*pi is ___
```

1.7 Introduction to Formatting

Fortran has extremely powerful, flexible, and easy-to-use capabilities for output formatting. This section describes the basic formatting features that enable you to produce really good-looking output, if you like. If the default formatting on your system is good enough, there is no necessity to learn formatting right away. This section appears early because some Fortran systems do not have satisfactory default formats, especially for reals. On such systems, the techniques of this section are essential.

Roundoff

Just as 1/3 cannot be represented exactly as a decimal, though 0.333333 comes very close, 1/10 and 1/100 cannot be represented exactly when the representation uses a number base two instead of ten. The base two or **binary system** of notation is used internally in most computers for storage and calculation of numeric values. As a result, when reals are converted from input represented in decimal notation to the computer's internal representation and back again during execution of a program, the original numbers may not be recovered precisely.

Perhaps you have already seen this in your own output, in the form of a tell-tale sequence of 9s in the decimal digits printed. For example, when adding 97.6 and –12.9 using the program `add_2_reals` in 1.3, the following output resulted.

```
Input data  x:  97.5999985
Input data  y: -12.8999996
x + y =  84.6999969
```

The value of the variable `x` prints as 97.5999985 although the value supplied in the input file is 97.6. The difference between the intended and calculated values is **round-off** or **roundoff error.** It is normally of no consequence in calculations of measured physical quantities because it is virtually impossible to distinguish between such nearly equal values as 97.5999985 and 97.6.

Similarly, the printed value of the variable `y` is –12.8999996 instead of –12.9. The printed value of `x + y` is 84.6999969 differing by 0.000031 from the sum of the intended values, a hint to the expert that the computer being used probably does not use decimal arithmetic for its internal calculations.

Using Formatted Output to Hide Roundoff

Minor cases of roundoff are hidden easily by rounding values before printing. For example, if the unexpected echoes of input data above are rounded to four decimal places before printing, the results will appear precisely as expected: 97.6000 + (–12.9000) = 84.7000.

If the default format for reals rounds answers to fewer decimal places than are actually calculated, you will not see any trace of roundoff. These extra guard digits may actually contain roundoff, but rounding answers before printing guarantees that the user will not see small roundoff errors. We mention roundoff at this point to forewarn the beginner whose system shows such behavior in output. Roundoff is not a malfunction of the computer's hardware, but a fact of life of finite precision arithmetic on computers. A programmer needs to know how to hide roundoff through formatted printing and needs to know why real values that print identically may still fail a test for equality.

In the remainder of this section we introduce the simplest forms of user-specified `print` formatting, including the facility for rounding real values to a specified number of decimal places before printing.

Format Specifications

Extremely flexible and versatile control over the appearance of printed output is available in Fortran if you are willing to forego the convenience of the default format. In place of the asterisk denoting the default format, write a format specification or some alternative means of locating a format specification in the program. A **format specification** is basically a list of **edit descriptors**, separated by commas and enclosed in parentheses. An example is

```
(f5.1, a, i4)
```

For each expression to be printed, one of the edit descriptors in the format specification is used to determine the form of the output. For example if `x` = 6.3 is type real and `n` = −26 is type integer, then

```
print "(f5.1, es9.1, a, i4)", x, x, " and ", n
```

would produce the output line

```
  6.3   6.3E+00 and  -26
```

This example shows four of the most frequently used edit descriptors, `f` (floating point) and `es` (engineering and science) for printing of reals, `a` (alphanumeric) for character strings, and `i` (integer) for integers. The edit descriptor `f5.1` means that a total of five positions are reserved for printing a real value rounded to one place after the decimal point. The decimal point occupies a position and a minus sign, if needed, occupies another position, so the largest number printable using `f5.1` format is 999.9 and −99.9 is the smallest. If the number to be printed is outside these bounds, the specified field will be filled with asterisks. `i4` editing reserves four positions for printing an integer. For negative numbers, the minus sign takes up one of the four positions, making `i4` format suitable for integers from −999 to 9999. The `es` (engineering/science) edit descriptor is used for printing reals in exponential notation. For example, the `es10.3` descriptor uses 10 positions, prints the most significant digit to the left of the decimal point, and prints the fractional part rounded to three decimal places, for example `6.023e+23` preceded by a blank character. For more details, see 11.8. The `a` edit descriptor reserves space for character output. The length of the character expression to be printed determines how many positions are used. It is also possible to reserve a specific number of positions for a character string. The edit descriptor `a10`, for example, reserves 10 positions, regardless of the data to be printed. See 11.8 for details.

Placement of Format Specifications

In the preceding example, the format specification is in the form of a character constant. Now the necessity of the comma after the asterisk or other format specifier in the `print` statement becomes apparent. It is the means of separating the format specifier from the first item in the list of expressions to be printed.

Since the format is a character expression, in the simplest case it is simply a character constant that appears in the input/output statement. For example, the following two `print` statements would produce the same output. It is assumed that `x` is real and `n` is integer.

```
character(len=*), parameter :: layout = "(f5.1, a, i4)"
print "(f5.1, a, i4)", x, " and ", n
print layout, x, " and ", n
```

Tab and Line Feed Edit Descriptors

The slash (/) edit descriptor starts a new line in the printed output. Thus, a single `print` statement can produce several lines of output. For example

```
print "(a, /, a, /, a)", "These character strings", &
      "all appear", "on separate lines."
```

produces the three lines of output

```
These character strings
all appear
on separate lines.
```

The `t` (tab) edit descriptor is used to skip to a specified position of the output line for precise control over the appearance of the output. Tabs may be either forward or backward on the current line. For example,

```
print "(t30, i5, t50, i5, t10, i5)", a, b, c
```

will print the integer values of `c` in positions 10–14, `a` in positions 30–34, and `b` in positions 50–54. Some printing devices do not print position 1 of any output line. If you have such a printer on your system, a `t2` edit descriptor will skip to position 2 to get single spacing.

Repeated Edit Descriptors

If one or more edit descriptors are to be repeated, they may be enclosed in parentheses and preceded by the positive integer representing the number of repetitions or an asterisk (*) representing an unlimited number of repetitions.

`3(i4)` is equivalent to `3i4` or `i4,i4,i4`
`5(/)` is equivalent to `5/` or `/,/,/,/,/`
`2(a4,/,t2)` is equivalent to `a4,/,t2,a4,/,t2`

The parentheses may be omitted if there is only one `a`, `es`, `f`, `i`, or `/` edit descriptor inside the parentheses.

Examples of Formatted Output

The following examples illustrate how formatted output works. On some printers, the first character may not appear, so it is best to put a blank in the first position.

```
print "(3i2)", 2, 3, 4

 2 3 4
```

```
x = 7.34688e-9
```

```
print "(a, es10.3)", " The answer is ", x
```

```
 The answer is  7.347E-09
```

```
q1 = 5.6
q2 = 5.73
q3 = 5.79
f123 = "(a, 3(/, t2, a, i1, a, f3.1))"
print f123, " Here come the answers--", &
      " q", 1, "=", q1, &
      " q", 2, "=", q2, &
      " q", 3, "=", q3
```

```
 Here come the answers--
  q1=5.6
  q2=5.7
  q3=5.8
```

Formatted Input

A format specification can be used with the **read** statement to indicate how the positions of the input line are to be interpreted. Formatted input is not as essential as formatted output because most natural arrangements of input data are accepted by the default **read** formats. However, there are two major exceptions, which sometimes make the use of input formatting desirable. First, default formats for character input usually require quotes around the input strings; character input read under an **a** edit descriptor does not. Second, it is a small convenience not to have to separate numbers with commas or blanks when large amounts of data are read by a program. For example, it is much harder to type 10 one-digit integers on a line of input with separating commas than without them. Rather than discuss the rules in detail for using formatted input, one example is given.

```
real :: x1, x2, x3
integer :: j1, j2, j3
character(len=4) :: c
read "(a, 3 (f2.1, i1))", c, x1, j1, x2, j2, x3, j3
```

If the input line is

```
1234567890123
```

then executing the **read** statement is equivalent to executing the following assignment statements. Notice that quotes for **a** format input data must be omitted and that decimal points for **f** format input data are assumed when they are omitted.

```
c = "1234"
x1 = 5.6
j1 = 7
```

```
x2 = 8.9
j2 = 0
x3 = 1.2
j3 = 3
```

Style note: It is good programming practice to use the default read format whenever possible. Explicit input format specifications demand strict adherence to specified positions for each value in the input data. The slightest misalignment of the input data usually results in incorrect values assigned to the variables. By comparison, the default input format is usually relatively tolerant of variations in alignment and is user-friendly.

Exercises

1. If the variable `x` has value 2.5, what does the output for the following statement look like? Use "*b*" for blank positions.

```
print "(f6.3, es11.1)", x, x ** 2
```

2. What are the largest and smallest values that can be printed by the statement

```
print "(f8.3)", value
```

3. What does the following statement print? Use "*b*" for blank positions.

```
print "(a, f9.5, a)", "!", 1.0/3.0, "!"
```

1.8 Case Study: Quadratic Formula

A quadratic equation is an equation involving the square of the unknown x and no higher powers of x. Algorithms for solution of quadratic equations equivalent to the quadratic formula are found in Old Babylonian texts dating to 1700 B.C. It is now routinely taught in high-school algebra. In this section, we show how to write a Fortran program to evaluate and print the roots of a quadratic equation. We also discuss improving the efficiency of the calculation by isolating common subexpressions. Sometimes there are better ways to solve a quadratic equation, particularly in cases where roundoff might be a problem (Exercise 2). Also the programs in this section do not handle the case when a, the coefficient of the x^2 term, is zero.

The Problem

The most general quadratic equation has the form

$$ax^2 + bx + c = 0$$

where a, b, and c are constants and x is the unknown. The quadratic formula says that the roots of the quadratic equation, that is, the values of x for which the equation is true, are given by the formula

$$x = \frac{-b \pm \sqrt{b^2 - 4ac}}{2a}$$

This means that one root is obtained by adding the square root term and the other root is obtained by subtracting the square root term.

The problem is to write a program that reads as input the three coefficients, a, b, and c, and prints as output the values of the two roots. Since there is very little input, and we wish to display the answers as they are computed, we write the program for interactive execution with input from a terminal keyboard and output to the display screen or printing element.

The Solution

Experienced programmers may regard the following pseudocode solution as obvious, as indeed it is, but the three steps of the pseudocode solution must be considered, if not necessarily written down.

```
Read the coefficients a, b, and c.
Calculate the two roots by the quadratic formula.
Print the two roots.
```

It is but a small step to the Fortran program that implements the pseudocode solution. Remember that an exclamation mark (!) begins a comment.

```
program quadratic_equation_solver
!  Calculates and prints the roots
!  of a quadratic equation
   implicit none
!  Variables:
!     a, b, c: coefficients
!     x1, x2: roots

   real :: a, b, c, x1, x2

!  Read the coefficients
   print *, "Enter a, the coefficient of x ** 2"
   read *, a
   print *, "Enter b, the coefficient of x"
   read *, b
   print *, "Enter c, the constant term"
   read *, c

!  Calculate the roots by the quadratic formula
   x1 = (-b + sqrt(b ** 2 - 4 * a * c)) / (2 * a)
```

```
   x2 = (-b - sqrt(b ** 2 - 4 * a * c)) / (2 * a)

!  Print the roots
   print *, "The roots are"
   print *, "x1 =", x1
   print *, "x2 =", x2
end program quadratic_equation_solver
```

In the input section, each `read` statement is preceded by an input prompt, that is, a `print` statement telling the user at the computer terminal what input is expected. In the calculation section, the quadratic formula illustrates the use of the intrinsic function `sqrt`.

Program Testing

To test the program `quadratic_equation_solver`, we made up several quadratic equations with known roots. Since all variables are type real, our first test case has simple real roots. The solutions of the quadratic equation

$$x^2 - 5x + 6 = 0$$

are 2 and 3.

```
Enter a, the coefficient of x ** 2
1
Enter b, the coefficient of x
-5
Enter c, the constant term
6
The roots are
x1 =    3.0000000
x2 =    2.0000000
```

The next quadratic equation has negative and fractional roots to test whether the program will work in these cases. The solutions of the quadratic equation

$$4x^2 + 8x - 21 = 0$$

are –3.5 and 1.5, testing both possibilities.

```
Enter a, the coefficient of x ** 2
4
Enter b, the coefficient of x
8
Enter c, the constant term
-21
The roots are
x1 =    1.5000000
x2 =   -3.5000000
```

Notice that `x1` is always the greater of the two roots because its formula adds the square root term.

The next case tests irrational roots of the quadratic equation. The golden ratio is a ratio famous from Greek mathematics. Renaissance artists thought that the golden ratio was the most pleasing ratio for the sides of a rectangular painting or the facade of a building. The spiral shells of snails and the arrangement of seeds in a sunflower are related to it. The golden ratio is also the limit of the ratio of successive terms of the Fibonacci sequence. The two roots of the following equation are the golden ratio and the negative of its reciprocal.

$$x^2 - x - 1 = 0$$

```
Enter a, the coefficient of x ** 2
1
Enter b, the coefficient of x
-1
Enter c, the constant term
-1
The roots are
x1 =    1.6180340
x2 =   -0.6180340
```

The exact solutions are $(1 + \sqrt{5}) / 2$ and $(1 - \sqrt{5}) / 2$, which check with the output of the program using a hand calculator. The golden ratio has many interesting properties, including the fact that 1/1.618034 = 0.618034.

The quadratic equation

$$x^2 - 6x + 9 = 0$$

has only one solution, $x = 3$. You might wonder what a program designed to find two roots will do with this equation.

```
Enter a, the coefficient of x ** 2
1
Enter b, the coefficient of x
-6
Enter c, the constant term
9
The roots are
x1 =    3.0000000
x2 =    3.0000000
```

Mathematicians call the solution of this quadratic equation a *double root*. For this equation, the quantity $b^2 - 4ac$ is zero, so it does not matter whether its square root is added or subtracted in the calculation of a root. The answer is the same for both roots.

Next we try the equation

$$x^2 - 1000001x + 1 = 0$$

Running the program produces the results

```
Enter a, the coefficient of x ** 2
1
Enter b, the coefficient of x
-1000001
Enter c, the constant term
1
The roots are
x1 =    1.0000010E+06
x2 =    0.0000000E+00
```

The smaller root is not accurate because b and $\sqrt{b^2 - 4ac}$ are nearly equal. Cancellation of the significant digits occurs during the subtraction leaving an answer severely contaminated by rounding errors. A way to cope with this situation is discussed in most introductory texts on numerical computation (Exercise 2).

Finally, we try a test case which we know the program quadratic_equation_solver cannot handle. The quadratic equation

$$x^2 + 1 = 0$$

has no real roots. Instead, the roots are

$$x = \pm\sqrt{-1} = \pm i$$

$+i$ and $-i$ are complex numbers with no real part. We still try it anyway, just to see what happens.

```
Enter a, the coefficient of x ** 2
1
Enter b, the coefficient of x
0
Enter c, the constant term
1
*** Attempt to take square root of negative quantity ***
*** Execution terminated ***
```

Since $b^2 - 4ac$ is -4, the error message is right on the money. One way to cope with this situation is discussed later.

Common Subexpressions

The arithmetic expressions for calculating the roots x1 and x2 both involve the same subexpression, sqrt(b**2 - 4*a*c). As written, the program quadratic_equation_solver asks the computer to recalculate this subexpression as part of the calculation of x2. We can force the computer to calculate this subexpression only once by assigning it to a new intermediate variable sub_expression, and then calculating both roots in terms of the variable sub_expression.

```
program quadratic_equation_solver_2
!  Calculates and prints the roots
!  of a quadratic equation
```

```
   implicit none
!  Variables:
!     a, b, c: coefficients
!     sub_expression: value common to both roots
!     x1, x2: roots

   real :: a, b, c, x1, x2, sub_expression

!  Read the coefficients
   print *, "Enter a, the coefficient of x ** 2"
   read *, a
   print *, "Enter b, the coefficient of x"
   read *, b
   print *, "Enter c, the constant term"
   read *, c

!  Calculate the roots by the quadratic formula
   sub_expression = sqrt (b ** 2 - 4 * a * c)
   x1 = (-b + sub_expression) / (2 * a)
   x2 = (-b - sub_expression) / (2 * a)

!  Print the roots
   print *, "The roots are"
   print *, "x1 =", x1
   print *, "x2 =", x2
end program quadratic_equation_solver_2
```

Some optimizing Fortran compilers will recognize that the program `quadratic_equation_solver`, in its original form, calls for the calculation of the same subexpression twice without change of any of the variables in the subexpression. Such a compiler would produce the more efficient machine language code corresponding to the second version, `quadratic_equation_solver_2`, even when the programmer writes the less efficient first version.

Complex Roots of a Quadratic Equation

The quadratic formula was used in the program `quadratic_equation_ solver` to calculate the roots of a quadratic equation. The program worked well when the two roots were real, but it failed in the test case of a quadratic whose roots were imaginary. In that case, the quadratic formula calls for taking the square root of a negative number, a function evaluation with no real answer. In the next program, `quadratic_equation_solver_3`, we use complex values to compute the correct answer whether the roots of the quadratic are real or complex.

The subexpression

$$d = b^2 - 4ac$$

is called the **discriminant** because it discriminates between the cases of two real roots, a double real root, and two complex roots. If d is positive, there is a real square root of d and the quadratic formula gives two real roots, one calculated by adding the square root of d and the other by subtracting it. If d is zero, so is its square root. Consequently, when d is zero the quadratic formula gives only one real root, $-b/2a$.

When d is negative, on the other hand, its square root is imaginary. The complex square root of a negative number is obtained by taking the square root of its absolute value and multiplying the result by i, the basis of the complex number system. For example, if $d = -4$, then $\sqrt{d} = 2i$. Thus when d is negative, the two roots of the quadratic equation are given by the formulas

$$x_1 = \frac{-b}{2a} + \frac{\sqrt{|d|}}{2a}i \quad \text{and} \quad x_2 = \frac{-b}{2a} - \frac{\sqrt{|d|}}{2a}i$$

However, with the use of the complex data type, the formula for calculating the roots looks just like it does when the roots are real. The only thing that makes quadratic_equation_solver_3 look different from the real version is that the discriminant is converted to a complex value and all the remaining computations are done with complex values. The two sample executions show one case where the roots are complex and one case where they are both real.

```
program quadratic_equation_solver_3

!  Calculates and prints the roots
!  of a quadratic formula even if they are complex

   implicit none
!  Variables:  a, b, c = coefficients
!              z1, z2 = roots

   real :: a, b, c
   complex :: z1, z2

!  Read the coefficients
   read *, a, b, c
   print *, "Input data  a:", a
   print *, "            b:", b
   print *, "            c:", c

!  Calculate the roots
   z1 = (-b + sqrt (cmplx (b**2 - 4*a*c))) / (2*a)
   z2 = (-b - sqrt (cmplx (b**2 - 4*a*c))) / (2*a)

!  Print the roots
   print *, "The roots are:"
   print *, "z1 =", z1
   print *, "z2 =", z2
```

```
end program quadratic_equation_solver_3

 Input data   a:    1.0000000
              b:    0.0000000E+00
              c:    1.0000000
 The roots are:
 z1 = (  0.0000000E+00,  1.0000000)
 z2 = (  0.0000000E+00, -1.0000000)

 Input data   a:    4.0000000
              b:    8.0000000
              c:  -21.0000000
 The roots are:
 z1 = (  1.5000000,  0.0000000E+00)
 z2 = ( -3.5000000,  0.0000000E+00)
```

Exercises

1. All of the programs in this section ignore the possibility that the value of **a** is zero, or is close to zero. What will happen if **quadratic_equation_solver** is run with input **a** = 0? Modify the program to handle this case. If **a** = 0, what happens if **b** is also zero? Modify the program to handle this case also. (Section 2.2 explains how to test if **a** = 0.)

2. If x_1 and x_2 are the roots of the quadratic equation $ax^2 + bx + c = 0$, their product is $x_1 x_2 = c/a$. When b is much larger than either a or c, the usual quadratic formula

$$x = \frac{-b \pm \sqrt{b^2 - 4ac}}{2a}$$

does a poor job of calculating the root with the smaller absolute value because the numerator is the difference of two nearly equal quantities $\sqrt{b^2 - 4ac}$ and b. Such subtractions always reduce the number of significant digits in the answer by the number of leading significant digits that cancel in the subtraction. Write a program that calculates the roots of a quadratic equation using the quadratic formula, and then recalculates the smaller root in absolute value using the formula $x_2 = c/ax_1$. Compare the two sets of roots. Test the program on the following equations.

$x^2 - 10x + 1 = 0$

$x^2 - 100x + 1 = 0$

$x^2 - 1000x + 1 = 0$

$x^2 - 10000x + 1 = 0$

1.9 Case Study: Debugging Pendulum Calculations

This section will explain some of the steps in making a real program work. The time it takes a pendulum to complete one swing is virtually independent of the amplitude or maximum displacement of the pendulum at the height of its swing, as long as the swing is relatively small compared with the length of the pendulum. For this reason, pendulums have long been used to keep accurate time. The problem in this section is to write a program to calculate the frequency f (the number of swings per second) of a pendulum, and its period T (the time it takes to complete one swing). The input data is the length of the pendulum in meters.

The formula for the frequency of a pendulum is

$$f = \frac{1}{2\pi}\sqrt{\frac{g}{L}}$$

where g is the gravitational acceleration constant 9.80665 m/s^2 for bodies falling under the influence of gravity near the surface of the Earth, L is the length of the pendulum in meters, and π is the mathematical constant 3.14159. In addition, the formula for the period T is

$$T = \frac{1}{f}$$

The First Compilation

The solution to this problem uses everything we learned in this chapter: It has variables, input data, computational formulas, and even the built-in square root function. Nevertheless, it seems to be a straightforward calculation for which a program can be written quite easily. Here is the first attempt.

```
program pendulum
!  Calculates the frequency and period
!  of a pendulum of length L

!  First attempt

   implicit none
   real :: L, f, T
   real, parameter :: pi = 3.14159,
                      g = 9.80665

   print *, "Enter a value for L: "
   read *, L
   f = (1.0 / 2.0 * pi) sqrt (g / L)
   T = 1.0 / f
end program pendulum
```

When this program is entered into the computer, it will not compile and run. The error messages we show below are illustrative approximations of the messages we get from actual Fortran compilers. The quality and amount of useful information contained in error messages varies widely. We suggest comparing the error messages shown here with the messages your system produces for the same errors.

```
% f03 pendulum.f03

Error: pendulum_1.f95, line 9: syntax error
       detected at ,@<end-of-statement>
***Invalid item in type declaration
Error: pendulum_1.f95, line 10: Implicit type for G
       detected at G@=
Error: pendulum_1.f95, line 14: syntax error
       detected at )@SQRT
***Malformed statement
```

Only three syntax errors is not too bad for a first attempt. The first error message is puzzling. What syntax error? The parameter assignment `pi = 3.14159` looks correct, echoed in the error message, and the parameter assignment `g = 9.80665` clearly is there in the next line. The second error message is even more puzzling. `g` is declared right in the line flagged by the error message. The crucial clue is before us, but as in a good detective mystery, only the practiced eye can see it. Looking back at the first error message, we now see that the compiler does not consider the line `g = 9.80665` to be a part of the `real` statement. Now the problem is clear. Both error messages are related, and both are caused by the same mistake. There is no continuation character (&) at the end of the first line of the statement declaring the parameters. It should read

```
real, parameter :: pi = 3.14159, &
                   g = 9.80665
```

The compiler sees the comma, and therefore expects another parameter assignment, but, in the absence of the continuation character, it finds the end-of-statement instead. Sometimes, when a compiler gets confused, it gets very confused. It would take a very clever compiler to consistently print a good error message such as

```
*** Error -- missing continuation character ***
```

The third error message said that the compiler was expecting something else when `sqrt` was found instead. The rule is that the asterisk for multiplication cannot be omitted in places where a multiplication sign can be omitted in ordinary algebraic notation. We correct this assignment statement to the following.

```
f = (1.0 / 2.0 * pi) * sqrt (g / L)
```

The Second Compilation and Run

Since all known errors have been corrected, we recompile the program. This time, there are no error messages.

```
program pendulum
!  Calculates the frequency and period
!  of a pendulum of length L

   implicit none
   real :: L, f, T
   real, parameter :: pi = 3.14159, &
                      g = 9.80665

   print *, "Enter a value for L:"
   read *, L
   f = (1.0 / 2.0 * pi) * sqrt(g / L)
   T = 1.0 / f
   print *, "The frequency of the pendulum is", &
            f, "swings/sec."
   print *, "Each swing takes", T, "sec."
end program pendulum
```

The input data should be the length of the pendulum in meters. Visualizing the size of a grandfather clock, and rounding the length of its pendulum to the nearest whole meter, we will use an input length of one meter. Here is what the second run produces.

```
 Enter a value for L:
1
The frequency of the pendulum is   4.9190345 swings/sec.
Each swing takes   0.2032919 sec.
```

The program does run to completion; it prints the answers, but they are wrong! The pendulum of a grandfather clock does not make almost five complete swings per second. One swing every two seconds is more like it, with each half of the swing producing a tick at one-second intervals. Just because the computer prints an answer, it does not necessarily mean that the answer is right. The computer's arithmetic is almost certainly perfect, but the formula it was told to compute might be in error.

All the evidence seems to be pointing a finger at the assignment statement to calculate the frequency f:

```
f = (1.0 / 2.0 * pi) * sqrt(g / L)
```

or, if that statement is correct, at the statements that assign values to the variables and parameters that appear on the right in that statement. The assignment statement for `f` seems at first glance to be the equivalent of the algebraic formula for the frequency, so we shift our attention to the assignment of the parameters `pi` and `g` and the reading of the variable `L`. The echo of input data shows that `L` is correct. The `parameter` statement assigning `pi` and `g` seems to be correct, so we shift our attention back to the assignment statement calculating `f`. The error must be in this statement. If we still do not believe that it is wrong, we could print the values of `pi` and `g` just before this statement to further narrow the focus.

Remember the rule that a sequence of multiplications and divisions is executed from left to right. Thus, the assignment statement executes as though it were written

```
f = ((1.0 / 2.0) * pi) * sqrt(g / L)
```

The correct version of the statement is

```
f = (1.0 / (2.0 * pi)) * sqrt(g / L)
```

The Third Compilation and Run

This time, the answers look correct. We expected a pendulum one meter long to swing once every two seconds.

```
Enter a value for L:
1
The frequency of the pendulum is   0.4984032 swings/sec.
Each swing takes   2.0064075 sec.
```

To check it, we calculate the algebraic formulas on a hand calculator and get the same answers, and we could also try other pendulum lengths in the computer.

Postmortem Discussion

The author is really not incompetent enough to make all of the errors shown in this 14-line program, at least not in one grand *tour de force*. However, even experienced programmers will make each of these errors, one at a time or in combination, over the course of writing several dozen longer programs. Thus, it is vital for programmers not only to know how to write programs, but also to have effective strategies for debugging programs when the inevitable bugs appear. The techniques illustrated above: compiler error messages, well-chosen test cases worked by hand, and diagnostic printed output will serve the programmer in good stead throughout a career. Some Fortran systems provide even more sophisticated tools.

2 Control Constructs

The programs in Chapter 1 performed simple calculations and printed the answers, but each statement in these programs was executed exactly once. Almost any useful program has the property that some collections of statements is executed many times, and different sequences of statements are executed depending on the values of the input data.

The Fortran statements that control which statements are executed, together with the statements executed, are called **control constructs**. Three control constructs, the `if` construct, the `case` construct, and the `do` construct are discussed in this chapter. Two related constructs, the `block` construct and the `associate` construct also are discussed. The `if`, `stop`, `go to`, and `continue` statements also are discussed briefly. Related topics are the `return` statement (3.12), masked array assignment (4.1), and the `select type` construct (12.4).

2.1 Statement Blocks

A collection of statements whose execution is controlled by one of the control constructs is called a **block**. For example, the statements between an `if` statement and the next matching `else if` statement form a block. Transferring control into a block from outside is not allowed, but it is possible to leave a `block` with a transfer of control. Any block may contain a complete `if`, `case`, `do`, or `block` construct, so that these constructs may be nested to any level.

The Block Construct

A block may be created by the **block construct** statements. This allows a variable to have a scope consisting of just such a block. For example, in the following statements, there are two variables named `x`. One is a type real parameter with the value 1.1; the other is type integer and is known only inside the block construct, where it is declared. Outside the block construct, `x` refers to the real parameter with value 1.1, as seen by the output of the program.

```
program block_test

   implicit none
   real, parameter :: x = 1.1
```

```
   block
      integer :: x
      do x = 1, 3
         print *, x
      end do
   end block

   print *, x

end program block_test
```

```
           1
           2
           3
   1.10000002
```

The *associate* Construct

The associate construct is a simple way to create an alias for a more complex expression, so that the alias may be used in the code, rather than rewriting the more complicated expression. Here are two simple examples. Note that in the first `associate` construct, changing the values of `x` and `y` does not change the value of `s`, because it is the alias of an expression that is not a variable, but in the second construct, `s` is the alias of the variable `x`, so changing `x` also changes `s`. Note also that `s` is not declared.

```
program assoc
   implicit none
   real :: x = 3, y = 4
   associate (s => sqrt(x**2 + y**2))
      print *, s
      x = 5; y = 12
      print *, s
   end associate
   associate (s => x)
      print *, s
      x = 9
      print *, s
   end associate
end program assoc
```

Running the program produces the following.

```
5.00000000
5.00000000
5.00000000
9.00000000
```

Indentation of Blocks

Indentation of the blocks of a construct improves the readability of a program. The subordinate placement of the controlled blocks visually reinforces the fact that their execution is conditional or controlled.

> *Style note:* The statements of each block of a construct should be indented some consistent number of spaces more than the statements that delimit the block.

Construct Names

A construct may have a **construct name** on its first statement. It consists of an ordinary Fortran name followed by a colon. The end statement that ends the construct must be followed by the same construct name. This permits more complete checking that constructs are nested properly and provides a means of exiting any construct or cycling (2.4) more than one level of nested loop.

The `exit` Statement

The **`exit` statement** causes termination of execution of a construct. If the keyword `exit` is not followed by the name of a construct, the innermost do construct is exited; if there is a construct name, that named construct (and all active constructs nested within it) is exited. Examples of the `exit` statement occur throughout this chapter.

2.2 The `if` Construct

The `if` construct is a simple and elegant decision construct that permits the selection of one of a number of blocks during execution of a program. The general form of an `if` construct is

```
if (logical expression) then
 block of statements
else if (logical expression) then
 block of statements
else if (logical expression) then
 block of statements
else if...
  .
  .
  .
else
 block of statements
end if
```

The else if and else statements and the blocks following them may be omitted. The end if statement must not be omitted. The if construct may have a construct name. Some simple examples follow.

```
if (a == b) then
   c = a
   print *, c
end if

dicey: if (dice <= 3 .or. dice == 12) then
   print *, "You lose!"
else if (dice == 7 .or. dice == 11) then
   print *, "You win!"
else
   print *, "You have to keep rolling until you get"
   print *, "either a 7 or a", dice
end if dicey
```

The if-then statement is executed by evaluating the logical expression. If it is true, the block of statements following it is executed. Execution of this block completes the execution of the entire if construct. If the logical expression is false, the next matching else if, else, or end if statement following the block is executed. The execution of an else if statement is exactly the same; the difference is that an if-then statement must begin an if construct and an else if statement must not. The else and end if statements merely serve to separate blocks in an if construct; their execution has no effect.

The effect of these rules is that the logical expressions in the if-then statement and the else if statements are tested until one is found to be true. Then the block following the statement containing that test is executed, which completes execution of the if construct. If all of the logical conditions are false, the block following the else statement is executed, if there is one.

Case Study: Escape Velocity of a Rocket

If a rocket or other object is projected directly upward from the surface of the Earth at a velocity v, it will reach a maximum height h above the center of the Earth given by the formula

$$h = \frac{R_E}{1 - v^2/2gR_E}$$

where R_E is the radius of the Earth (6.366×10^6 m) and g is the acceleration due to gravity at the surface of the Earth (9.80 m/s^2). This formula is not an unreasonable approximation, since a rocket reaches its maximum velocity within a relatively short period of time after launching, and most of the air resistance is confined to a narrow layer near the surface of the Earth.

A close examination of this formula reveals that it cannot possibly hold for all velocities. For example, if the initial velocity v is such that $v^2 = 2gR_E$, then $1 - v^2/2gR_E$ is zero and the maximum height h is infinite. This velocity $v = 1.117 \times 10^4$ m/s (approximately 7 miles/s) is called the **escape velocity** of the Earth. Any object, either rocket or atmospheric gas molecule, attaining this vertical velocity near the surface of the Earth will leave the Earth's gravitational field and not return. A particle starting at the escape velocity will continue rising to arbitrarily great heights above the Earth. As it does so, it will slow to almost, but not quite, zero velocity.

At initial velocities greater than the escape velocity, the particle or rocket's velocity will not drop to zero. Instead it will escape from the Earth's gravitational field with a final velocity v_{final} given by the formula

$$v_{final} = \sqrt{v^2 - 2gR_E}$$

The original formula for the maximum height h gives negative answers in these cases and should not be used. The maximum height is infinite.

The Problem

We wish to write a program that reads an initial velocity of a rocket or molecule (in meters per second) and prints an appropriate description of the fate of the rocket or molecule. That is, if the rocket reaches a maximum height before falling back to Earth, the maximum height should be printed. On the other hand, if the rocket escapes the Earth's gravitational field, the final velocity with which it escapes should be printed.

The Solution in Pseudocode

From the preceding discussion, we see that the fate of the rocket or molecule can be determined by comparing the initial velocity to the escape velocity of the Earth, or equivalently, by comparing v^2 to $2gR_E$. If v^2 is smaller, then a maximum height h is reached before the rocket or molecule falls back to Earth. If the initial velocity is greater, then the object in question escapes with a nonzero final velocity given by the second formula. In the pseudocode solution below, the control structure is modeled exactly on the `if` construct.

```
Read the initial velocity v
Echo the input data
If (v^2 < 2gR_E) then
   Calculate maximum height h above center of Earth
   Print that the object attains maximum height h - R_E
      above the surface of the Earth before returning
      to Earth
else if (v^2 == 2gR_E) then
   Print that the initial velocity is
         the escape velocity
else
   Calculate the final velocity
```

```
      Print that the object escapes Earth
            with the calculated final velocity
   end if
```

The `if` construct extends from the keyword `if` that begins the `if` construct to the keyword `end if` that ends the construct. The two lines of pseudocode between the keyword `then` and the keywords `else if` constitute a block. They are executed if and only if $v^2 < 2gR_E$. The line of pseudocode between the second keyword `then` and the keyword `else` is the first and only block controlled by an `else if` statement in this `if` construct. It is executed whenever $v^2 = 2gR_E$. Finally, the two lines of pseudocode between the keyword `else` and the keyword `end if` are the `else` block. They are executed in case none of the preceding `if` or `else if` conditions are true.

The Fortran Solution

Little remains to be done to refine the pseudocode solution to an executable Fortran program except to choose names for the Fortran variables and parameters that most nearly resemble the variable names in the formulas and to translate the pseudocode to Fortran nearly line by line.

```
program escape
!  Accepts as input an initial velocity v
!  Prints maximum height attained,
!     if object does not escape Earth
!  Prints final escape velocity, vfinal,
!     if object escapes

!  Parameters
!     g  = acceleration of gravity near Earth's surface
!             in meters / sec ** 2  (m/s**2)
!     RE = radius of the Earth (in meters)

   implicit none
   real :: v, h, vfinal
   real, parameter :: g = 9.80, RE = 6.366e6

   read *, v
   print *, "Initial velocity of object =", v, "m/s"
   if (v ** 2 < 2 * g * RE) then
      h = RE / (1 - v ** 2 / (2 * g * RE))
      print *, "The object attains a height of",  &
               h - RE, "m"
      print *, "above the Earth's surface " //  &
               "before returning to Earth."
   else if (v ** 2 == 2 * g * RE) then
      print *, "This velocity is the escape " //  &
               "velocity of the Earth."
      print *, "The object just barely escapes " //  &
               "from Earth's gravity."
```

```
      else
         vfinal = sqrt (v ** 2 - 2 * g * RE)
         print *, "The object escapes with velocity", &
                  vfinal, "m/s."
      end if
   end program escape
```

```
 Initial velocity of object =   1.0000000E+03 m/s
 The object attains a height of   5.1432500E+04 m
 above the Earth's surface before returning to Earth.

 Initial velocity of object =   2.0000000E+04 m/s
 The object escapes with velocity   1.6589949E+04 m/s.

 Initial velocity of object =   1.1170000E+04 m/s
 The object attains a height of   1.6871994E+11 m
 above the Earth's surface before returning to Earth.
```

Testing an `if` Construct

The goal in testing an `if` construct is to design test cases that exercise each alternative in the `if` construct. The first sample execution shows an initial velocity of 1.0×10^3 m/s (1 km/s), which is well below the escape velocity of the Earth. The sample execution shows that the rocket reaches a maximum height of 5.14×10^4 m (51.4 km) before falling back to Earth. Calculating the appropriate formula using a hand calculator gives the same answer.

The second sample execution shows an initial velocity of 2.0×10^4 m/s (20 km/s), which is well above the escape velocity. As expected, the printed output shows that the rocket will escape from the Earth's gravitational field, so the correct block in the `if` construct is executed. It may seem surprising at first that the final velocity on escape is such a large fraction of the initial velocity. We rechecked it using a hand calculator and got the same answer. The explanation is that an initial velocity of nearly twice the escape velocity carries with it an initial kinetic energy (energy of motion) of nearly four times the energy of the escape velocity. So it is not really surprising that nearly three-fourths of the initial kinetic energy is retained and carried away with the rocket in the form of a large final velocity.

The third sample execution is designed to test the program using the escape velocity 1.117×10^4 m/s (11.17 km/s) as the initial velocity. Unfortunately, there is a little bit of roundoff in the calculations, and the middle block in the if construct is not executed. The printed answer is not bad. It says that the rocket will rise to a height of 1.69×10^{11} meters above the surface of the Earth before returning. Since this height is farther than the distance to either Mars or Venus at their nearest approach to Earth, for all practical purposes the program has reported that the rocket will escape.

Roundoff Error in Tests for Equality

You must expect some roundoff in any calculation using reals. The largest source of roundoff in this problem is the fact that the physical constants, the radius of the Earth, and the gravitational acceleration, are given to only three or four significant digits, as is the escape velocity. Even if the physical constants were given and used to more digits, each arithmetic calculation in the computer is calculated to a fixed number of digits. If you run this program on your computer, you will probably notice that the last one or more digits of your computer's printed answers differ from the ones shown. This is to be expected. We suggest that you try initial velocities slightly larger than 1.117×10^4 m/s in an attempt to hit the escape velocity exactly on the nose. Quite likely there is no computer-representable number on your machine to use as input to cause execution of the middle alternative in the `if` block. *Equality tests for reals are satisfied only in special circumstances.* The best you can reasonably expect is even larger maximum heights or extremely low final escaping velocities. To avoid this test for equality, test for *approximate equality* instead. In our case, the two values v^2 and $2gR_E$ probably should be considered equal if they agree to within three significant digits because `g` is given to only three significant digits. This test for approximate equality can be used to replace the `else if` statement in the program `escape`.

```
else if (abs((v**2 - 2*g*RE) / (2*g*RE)) < 1.0e-3) then
```

Flowchart for an `if` Construct

In standard flowcharting conventions, a diamond-shaped box is used to indicate a decision or fork in the flow of the program execution. A rectangular box represents processing of some sort. Using these standard conventions, the flowchart in Figure 2-1 indicates how an `if` construct is executed.

Case Study: Graduated Income Tax

The U.S. federal income tax is an example of a graduated or progressive tax, which means that each income level is taxed at a different rate. After all deductions, progressively higher incomes are taxed at increasing rates. A program to calculate federal income tax uses a multi-alternative `if` block to select the correct tax computation formula for each income level.

The resulting program illustrates the use of some of the logical operators `.and.`, `.or.`, and `.not.` To calculate a person's income tax liability, income for the year is modified by various exclusions, deductions, and adjustments to arrive at a taxable income. The problem treated in this section is that of writing a program to compute the

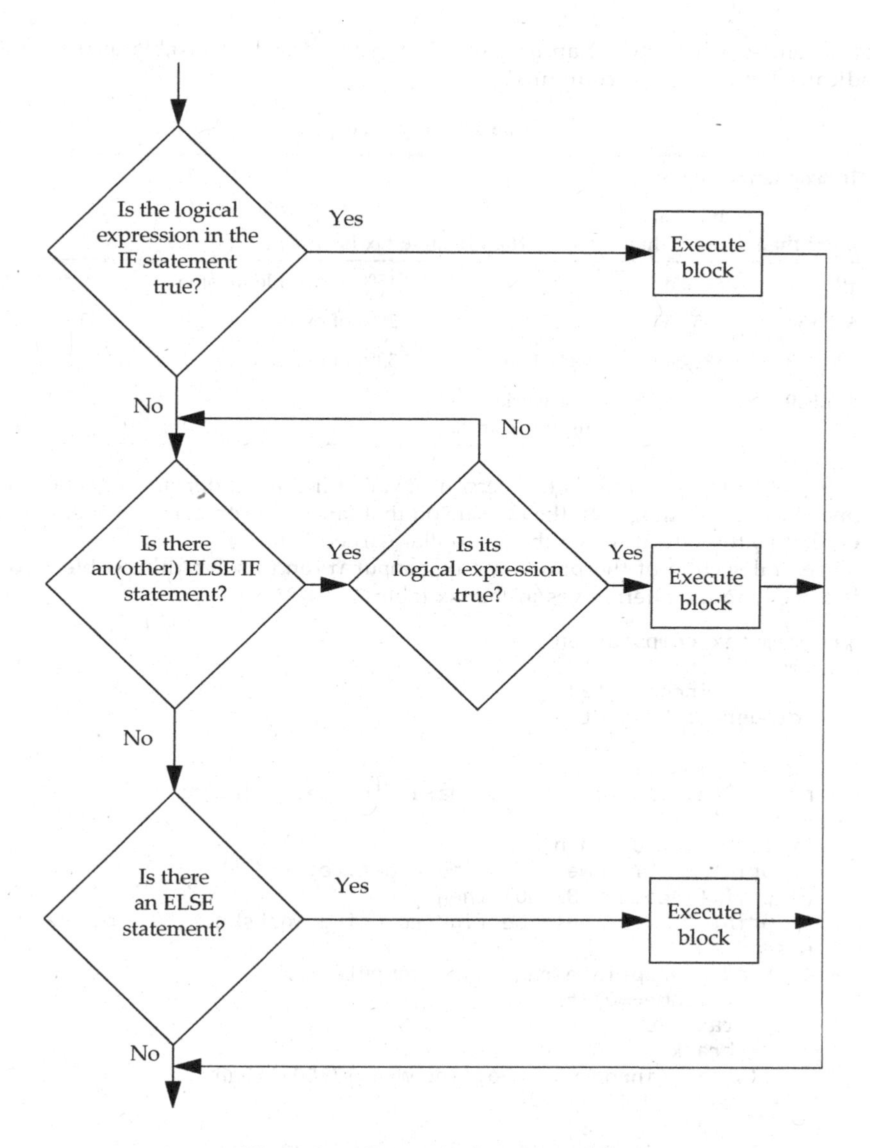

Figure 2-1 Execution flow for an `if` construct

federal income tax liability for an unmarried taxpayer based on taxable income. Table 2-1 indicates how the tax is computed.

Table 2-1 Tax table

If taxable income is				
more than	**but not more than**	**then income tax is**		
$0	$17,850		15% of taxable income	
$17,850	$43,150	$2,677.50 plus	28% of excess over	$17,850
$43,150	$81,560	$9,761.50 plus	33% of excess over	$43,150
$81,560	. . .	Use worksheet to figure your tax		

The input to the program is the person's taxable income, after all deductions and adjustments. The output is both the tax due on that taxable income and the person's tax bracket, that is, the rate at which the last dollar earned is taxed.

The central section of the program `tax_computation` to solve this problem corresponds directly to the alternatives in the tax table.

```
program tax_computation
   implicit none
   real :: income, tax
   integer :: bracket

   read *, income
   print "(a, f15.2)", "Input data  income:", income

   if (income < 0) then
      print *, "Income cannot be negative."
   else if (income > 81560) then
      print *, "Tax must be figured using worksheet."
   else
   !  Find appropriate range and compute tax
      if (income==0) then
         tax = 0
         bracket = 0
      else if (income>0 .and. income<=17850) then
         tax = 0.15 * income
         bracket = 15
      else if (income>17850 .and. income<=43150) then
         tax = 2677.50 + 0.28 * (income - 17850)
         bracket = 28
      else if (income>43150 .and. income<=81560) then
         tax = 9761.50 + 0.33 * (income - 43150)
         bracket = 33
```

```
      end if
   !  End of tax computation section
      print "(a, f8.2, a, f8.2)", &
            "The tax on $", income, " is $", tax
      print "(a, i2, a)", "This income is in the ",  &
            bracket, "% tax bracket."
   end if

end program tax_computation
```

Each line in the tax table corresponds to an if or else if test and a corresponding block in the tax computation if construct. If income lies in the indicated range for that if test, then the variables tax and bracket are calculated by the formula in the following block. The conditions describing the ranges for income follow the tax table exactly. They guarantee that only one range and one tax computation formula applies for each possible value of income less than or equal to $81,560.

To be more specific, let us look at a few sample executions of tax_computation, in which the computer is supplied with different values as input for the variable income.

```
Input data  income:        1000.00
The tax on $ 1000.00 is $  150.00
This income is in the 15% tax bracket.

Input data  income:       20850.00
The tax on $20850.00 is $ 3517.50
This income is in the 28% tax bracket.

Input data  income:       63150.00
The tax on $63150.00 is $16361.50
This income is in the 33% tax bracket.

Input data  income:       95000.00
 Tax must be figured using worksheet.
```

Consider the second run with a taxable income of $20,850. The only condition in the tax computation section that this taxable income satisfies is

```
income > 17850 .and. income <= 43150
```

The tax is computed by the formula in the following block

```
tax = 2677.50 + 0.28 * (income - 17850)
```

so that the tax computed is $2677.50 + 0.28 \times (20850 - 17850) = 2677.50 + 0.28 \times 3000 = 2677.50 + 840 = 3517.50$. The second assignment statement of this block assigns a tax bracket of 28 (percent) to the variable bracket. The remaining else if test in the if construct is skipped. Then the two print statements that complete the else block of the outer if construct are executed. Note that a complete if construct may be part of a block controlled by another if construct.

In the last sample execution, using a taxable income of $95,000, the condition in the else if statement of the outer if construct is satisfied, the variables tax and bracket are not assigned values at all, and the computer prints a statement that the tax cannot be computed using the tax table.

> *Style note:* It is good programming practice to warn the user when a situation occurs that the program is not designed to handle.

Nonexclusive `if` Conditions

Because the tax computation `if` construct in the program `tax_computation` is based so closely on the tax table, the alternative `if` and `else if` conditions are mutually exclusive. Just as one and only one line of the tax table applies to each taxable income, one and only one condition in the tax computation `if` construct is true (up to $81,560).

The test conditions in an `if` construct need not be mutually exclusive. Fortran permits more than one condition to be true. However, even if several conditions are true, only the first such condition selects its block for execution. The remaining conditions are not even tested. Executing the selected block completes execution of the entire `if` construct.

Using this rule for breaking ties when several conditions are satisfied, we may rewrite the inner `if` block of the program `tax_computation` with shorter test conditions.

```
!  Find appropriate range and compute tax
   if (income == 0) then
      tax = 0
      bracket = 0
   else if (income <= 17850) then
      tax = 0.15 * income
      bracket = 15
   else if (income <= 43150) then
      tax = 2677.50 + 0.28 * (income - 17850)
      bracket = 28
   else
      tax = 9761.50 + 0.33 * (income - 43150)
      bracket = 33
   end if
!  End of tax computation section
```

What is to be gained by shortening the `if` tests? Certainly, there is less typing to enter the program. In addition, since the `if` tests are simpler, they will execute more rapidly. Just how much more rapidly is not clear. Not only is the correspondence between the length of the Fortran source program and the speed of execution of the compiled machine language program rather loose, but input and output operations tend to be very time consuming when compared with computational statements. Thus, it is possible that most of the execution time is spent in the `read` and `print` statements, and even a significant improvement in the speed of the `if` tests produces very little change in the total execution time.

What is lost? The most important thing that is lost is the closeness of the correspondence between the program and the tax table. The original program tax_computation obviously implements the tax table, but although the new program also would implement the tax table, this fact would not be so obvious.

Another difference is that the second if construct is slightly more fragile or less robust. This means that although it works perfectly in its present form, it is slightly more likely to fail if it is modified at a later date. For example, if the order of the alternatives in the program tax_computation is scrambled, perhaps listed in decreasing rather than in increasing order of taxable income, the tax computation if construct in the original tax_computation program still works properly, but the replacement does not. The alternatives in the replacement if construct must remain in increasing order or the if construct will fail to compute taxes properly. On balance, the slight gain in efficiency and the slightly fewer keystrokes needed do not justify the less robust program.

> *Style note:* Do not sacrifice clarity of the program to shorten the execution time by a few nanoseconds. Not only is the program harder to get right and maintain, but with a good optimizing compiler the improvement in execution time may be smaller than anticipated or even nonexistent.

The if Statement

There is a special form of test that is useful when there are no else if or else conditions and the action to be taken when the condition is true consists of just one statement. It is the **if statement**. The general form of an if statement is

if (*logical expression*) *statement*

The statement to be executed when the logical expression is true must not be anything that does not make sense alone, such as an end if statement. Also, it must not be another if statement, and it may not be the first statement of a control construct.

When the if statement is executed, the logical expression is evaluated. If the result is true, the statement following the logical expression is executed; otherwise, it is not executed.

Using this form of testing has the drawback that if the program is modified in such a way that the single statement if is no longer adequate, the if statement must be changed to an if construct. If an if construct were used in the first place, the modification would consist of simply adding more statements between the if-then statement that begins the if construct and the end if statement. However, in the cases where the computation to be done when a certain condition is true consists of just one short statement, using the if statement probably makes it a little easier to read. Compare, for example

```
if (value < 0) value = 0
```

with

```
if (value < 0) then
   value = 0
end if
```

Exercises

1. Write an `if` construct that prints the word "vowel" if the value of the variable `letter` is a vowel (i.e., A, E, I, O, or U) and the word "consonant" if the value of `letter` is any other letter of the alphabet. Only uppercase letters can appear as values of `letter`. Hint: to test if the value of letter is A, write

```
if (letter == "A") then
```

2. Hand simulate the programs `example_1` to `example_4` using the values 45, 75, and 95 as input data (12 simulations in all). Check your answers with a computer, if possible. *Caution*: These simulations are tricky, but each program is syntactically correct. No indentation has been used in order not to give any hints about the structure of the `if` constructs. We suggest correctly indenting each program before hand simulating it.

```
program example_1
   integer :: x
   read *, x
   if (x > 50) then
   if (x > 90) then
   print *, x, " is very high."
   else
   print *, x, " is high."
   end if
   end if
end program example_1
```

```
program example_2
   integer :: x
   read *, x
   if (x > 50) then
   if (x > 90) then
   print *, x, " is very high."
   else
   end if
   print *, x, " is high."
   end if
end program example_2
```

```
program example_3
   integer :: x
   read *, x
   if (x > 50) then
   if (x > 90) then
   print *, x, " is very high."
   end if
   else
   print *, x, " is high."
   end if
```

```
end program example_3

program example_4
   integer :: x
   read *, x
   if (x > 50) then
   end if
   if (x > 90) then
   print *, x, " is very high."
   else
   print *, x, " is high."
   end if
end program example_4
```

3. A toll bridge charges $3.00 for passenger cars, $4.00 for buses, $6.00 for trucks under 10,000 pounds, and $10.00 for trucks over 10,000 pounds. The problem is to write a program using an `if` construct to compute the toll. Use interactive input if it is available. The input data consists of first the letter C, B, or T for car, bus, or truck, respectively. Either uppercase or lowercase letters are permitted. If the class is T (truck), then prompt the user for another character that is either "<" (meaning less than 10,000 pounds) or ">" (meaning greater than 10,000 pounds). The following are sample executions:

```
  Enter vehicle class (C, B, or T)
t
  Enter < or > to indicate weight class
<
  The toll is $6.00

  Enter vehicle class (C, B, or T)
c
  The toll is $3.00
```

   Hint: the vehicle class variable might be declared as

```
character(len=1) :: vehicle_class
```

4. The Enlightened Corporation is pleased when its employees enroll in college classes. It offers them an 80% rebate on the first $500 of tuition, a 60% rebate on the second $400, and a 40% rebate on the next $300. The problem is to compute the amount of the rebate. The input data consists of one number, the amount of tuition paid by the employee. A sample execution might produce the following:

```
Input data  tuition:  600
The employee's rebate is $  460
```

2.3 The case Construct

The `case` construct is somewhat similar to the `if` construct in that it permits selection of one of a number of different alternative blocks of instructions, providing a streamlined syntax for an important special case of a multiway selection. The general form of a case construct is

```
select case (expression)
 case (case selector)
   block of statements
 case (case selector)
   block of statements
   .
   .
   .
 [case default
   block of statements ]
end select
```

The value of the expression in the `select case` statement should be an integer or a character string (of any length). The case selector in each `case` statement is a list of items, where each item is either a single constant or a range of the same type as the expression in the `select case` statement. A **range** is two constants separated by a colon and stands for all the values between and including the two values. The `case default` statement and its block are optional. The `case` construct may have a construct name.

The `case` construct is executed by evaluating the expression in the `select case` statement. Then the expressions in the `case` statements are examined until one is found with a value or range that includes the value of the expression. The block of statements following this `case` statement is executed, completing execution of the entire `case` construct. Unlike `if` constructs, no more than one `case` statement may match the value of the expression. If no `case` statement matches the value of the expression and there is a `case default` statement, the block following the `case default` statement is executed.

Any of the items in the list of values in the `case` statement may be a range of values, indicated by the lower bound and upper bound separated by a colon (`:`). The case expression matches this item if the value of the expression is greater than or equal to the lower bound and less than or equal to the upper bound.

A flowchart indicating how a `case` construct is executed appears in Figure 2-2. Some simple examples follow.

```
select case (dice)
   case (2:3, 12)
      print *, "You lose!"
   case (7, 11)
      print *, "You win!"
```

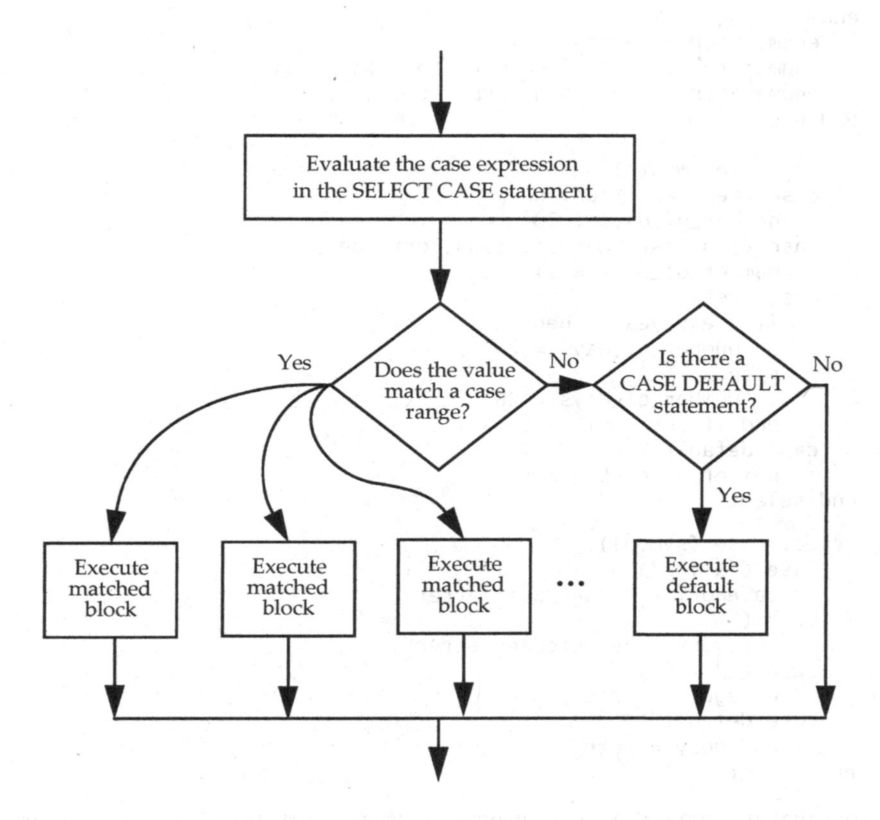

Figure 2-2 Execution flow for a case construct

```
   case default
      print *, "You have to keep rolling until you get"
      print *, "either a 7 or a ", dice
end select

traffic: select case (traffic_light)
   case ("red")
      print *, "Stop"
   case ("yellow")
      print *, "Caution"
   case ("green")
      print *, "Go"
   case default
      print *, "Illegal value:", traffic_light
end select traffic
```

```
enum, bind(c)
   enumerator :: error
   enumerator :: jan, feb, mar, apr, may, jun
   enumerator :: jul, aug, sep, oct, nov, dec
end enum
   . . .
select case (month)
   case (sep, apr, jun, nov)
      number_of_days = 30
   case (jan, mar, may, jul:aug, oct, dec)
      number_of_days = 31
   case (feb)
      if (leap_year) then
         number_of_days = 29
      else
         number_of_days = 28
      end if
   case default
      number_of_days = error
end select

select case (symbol)
   case ("a":"z")
      category = "lowercase letter"
   case ("A":"Z")
      category = "uppercase letter"
   case ("0":"9")
      category = "digit"
   case default
      category = "other"
end select
```

Note that the computation of income tax that was done in the previous section with an `if` construct cannot be done with a `case` construct because the data type of the expression used in a `select case` statement may not be real.

Exercises

1. Write a case construct that prints the word "vowel" if the value of the variable `letter` is a vowel (i.e., A, E, I, O, or U), prints the word "consonant" if the value of `letter` is any other letter of the alphabet, and prints an error message if it is any other character.

2. Write a complete program that reads one character and uses the `case` construct to print the appropriate classification of the character.

3. A toll bridge charges $3.00 for passenger cars, $2.00 for buses, $6.00 for trucks under 10,000 pounds, and $10.00 for trucks over 10,000 pounds. The problem is to write a program to compute the toll using a `case` construct. Use interactive input if

it is available. The input data consists of first the letter C, B, or T for car, bus, or truck, respectively. Either uppercase or lowercase letters are permitted. If the class is T (truck), then prompt the user for another character that is either "<" (meaning less than 10,000 pounds) or ">" (meaning greater than 10,000 pounds). The following are sample executions:

```
    Enter vehicle class (C, B, or T)
t
    Enter < or > to indicate weight class
<
    The toll is $6.00

    Enter vehicle class (C, B, or T)
c
    the toll is $3.00
```

2.4 The do Construct

All the programs so far suffer from the defect that each instruction is executed at most once. At the enormous speed at which computers execute instructions, it would be difficult to keep a computer busy for very long using this type of program. By the simple expedient of having the computer execute some instructions more than once, perhaps a large number of times, it is possible to produce a computer program that takes longer to execute than to write. More important is the fact that a loop increases the difficulty of writing a program very little, while it greatly increases the amount of useful data processing and calculation done by the program.

The looping construct in Fortran is the **do construct**. The general form of the do construct is

do [*loop control*]
  block of statements
end do

The do construct may have a construct name.

The block of statements, called the **loop body** or do construct body, is executed repeatedly as indicated by the loop control. Figure 2-3 is a flowchart showing the execution of a do construct.

There are two types of loop control. In one case the loop control is missing, in which case the loop is executed until some explicit instruction in the do body such as an `exit` statement terminates the loop. In the other type of loop control, a variable takes on a progression of values until some limit is reached. After a brief discussion of the `cycle` statement, we will look at examples of the different types of loop control.

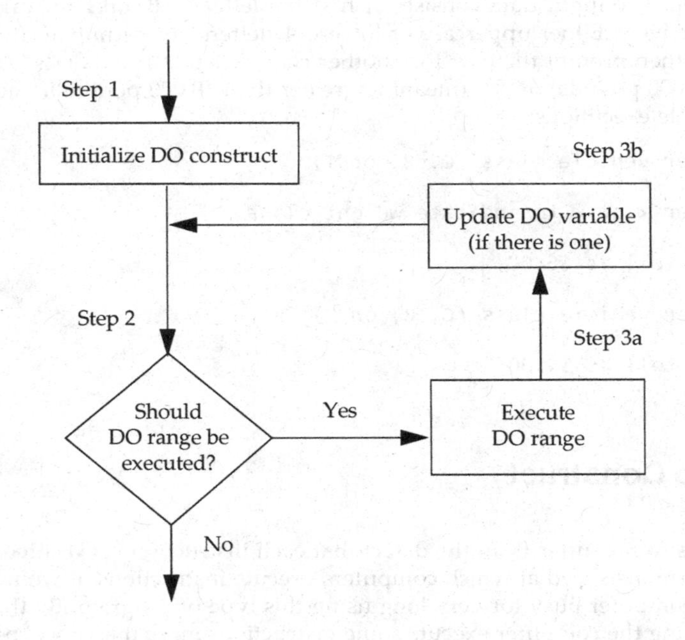

Figure 2-3 Execution flow for a *do* construct with an iteration count

The `cycle` Statement

The **`cycle` statement** causes termination of the execution of *one iteration* of a loop. In other words, the **do** body is terminated, the **do** variable (if present) is updated, and control is transferred back to the beginning of the block of statements that comprise the **do** body. If the keyword `cycle` is followed by the name of a construct, all active loops nested within that named loop are exited and control is transferred back to the beginning of the block of statements that comprise the named **do** construct.

Loops with No Loop Control

For a **do** construct with no loop control, the block of statements between the **do** statement and the matching **end do** statement are executed repeatedly until an `exit` statement or some other statement causes it to terminate. Suppose we wish to print out all powers of two that are less than 1000. This is done with a simple **do** construct with no loop control and an `exit` statement.

```
program some_powers_of_2

   implicit none
   integer :: power_of_2

   power_of_2 = 1  ! The zero power of 2
   print_power: do
      print *, power_of_2
      power_of_2 = 2 * power_of_2
      if (power_of_2 >= 1000) exit print_power
   end do print_power
end program some_powers_of_2
```

As another example, suppose a file contains integers, one per line. All of the integers are nonnegative, except the last integer in the file, which is negative. The following program reads the file and computes the average of the integers, treating the first negative integer it finds as a signal that there is no more data.

```
program average
!  This program finds the average of a file of
!  nonnegative integers, which occur one per line
!  in the input file. The first negative number
!  is treated as the end of data.

   implicit none
   integer :: number, number_of_numbers, total

   total = 0
   number_of_numbers = 0
   do
      read *, number
      if (number < 0) exit
      print *, "Input data  number:", number
      total = total + number
      number_of_numbers = number_of_numbers + 1
   end do

   print *, "The average of the numbers is", &
         real(total)  / number_of_numbers
end program average
```

To illustrate a simple use of the `cycle` statement, suppose a file of integers similar to the one used above is presented and the task is to count the number of odd numbers in the file prior to the first negative number in the file. The following program accomplishes this. Recall that the intrinsic function `modulo` gives the remainder when the first number is divided by the second number.

```
program odd_numbers
!  This program counts the number of odd numbers
!  in a file of nonnegative integers,
!  which occur one per line in the input file.
!  The first negative number is treated as end of data.

   implicit none
   integer :: number, number_of_odd_numbers

   number_of_odd_numbers = 0
   do
      read *, number
      print *, "Input data  number:", number
      if (number < 0) then
         exit
      else if (modulo (number, 2) == 0) then
         cycle
      else
         number_of_odd_numbers = number_of_odd_numbers + 1
      end if
   end do

   print *, "The number of odd numbers is", number_of_odd_numbers
end program odd_numbers
```

These last two programs have a structure similar to that of the heart of many programs, both simple and complicated. In pseudocode, that structure is

```
do
   Attempt to read some data
   If all data have been processed, then exit
   Process the data
end do
```

For this kind of loop, a do construct with no loop control and an exit statement are just right.

Loop Control with a do Variable

Quite frequently, the successive values taken by a variable follow a simple pattern, like 1, 2, 3, 4, 5, 6, 7, 8, 9, 10, or 9, 7, 5, 3. Because these sequences occur so often in programming, there is a simple means of assigning successive values to a variable in Fortran using the do construct and variable loop control. A simple example that prints the squares and cubes of the integers 1–20 follows:

```
do number = 1, 20
   print *, number, number**2, number**3
end do
```

The block of this do construct consists of a single print statement. The first time the print statement is executed, the do **variable** number has the value of 1, and this number is printed as the first output line, followed by its square and its cube. Then the do variable number takes on the value 2, which is printed on the next line, followed by its square and its cube. Then the do variable takes on the values 3, 4, 5, up to 20 for successive repetitions of the print statement. At this point, the possible values for the do variable number specified in the do statement are exhausted and execution of the do construct terminates.

The general forms of loop control using a do variable are

variable = *expression* , *expression*

and

variable = *expression* , *expression* , *expression*

The three integer expressions specify the starting value, the stopping value, and the step size or **stride** between successive values of the do variable. The do statement in the do construct above used constants 1 and 20 for the starting and stopping values. When the step size expression is omitted, as it is in the do construct above, a step size of 1 is used.

A do variable must be an integer variable; it should be declared in the program or procedure (3.13) where it is used. It must not be an array element (4.1) or a component of a structure (6.1). It should not be a dummy argument. It must not have the pointer or target attribute (10.1).

The value of a do variable may not be changed inside the construct.

The number of times the loop is executed (unless terminated by an exit statement, for example) is given by the formula

$$max\left(\left\lfloor \frac{m_2 - m_1 + m_3}{m_3} \right\rfloor, 0\right)$$

where m_1 is the starting value, m_2 is the stopping value, and m_3 is the step size. $\lfloor x \rfloor$ denotes the floor function, the greatest integer less than or equal to x. In cases where the sequence of values starting at m_1 in steps of m_3 exactly reaches m_2, this reduces to the simpler formula

$$1 + \frac{m_2 - m_1}{m_3}$$

For example, the following do loop is executed $\lfloor (10 - 2 + 2)/2) \rfloor = 5$ times with the do variable assigned the values 2, 4, 6, 8, and 10.

```
do number = 2, 10, 2
   print *, number
end do
```

If the do statement were changed to

```
do number = 2, 11, 2
```

The do loop would be executed $\lfloor (11 - 2 + 2)/2 \rfloor = 5$ times, as before, and the values of the do variable number would be the same: 2, 4, 6, 8, and 10. The do statement

```
do number = 1, upper_limit
```

causes its do block to be executed no times if the value of the variable upper_limit is less than or equal to zero.

Counting Backward

If the step size is negative, the do variable counts backwards. Thus, it is possible to print the complete words to the popular camp song "Ninety-Nine Bottles of Beer on the Wall" using a do statement with a negative step size. The program beer, which tells the computer to print the verses, is given below. In the program, a print statement with no print list is used to print a blank line between verses.

```
program beer
!  Prints the words of a camp song

   implicit none
   integer :: n

   do n = 99, 1, -1
      print *
      print *, n, "bottles of beer on the wall."
      print *, n, "bottles of beer."
      print *, "If one of those bottles should happen to fall,"
      print *, "there'd be", n - 1, "bottles of beer on the wall."
   end do
end program beer
```

Running the program produces the following output.

```
99 bottles of beer on the wall.
99 bottles of beer.
If one of those bottles should happen to fall,
there'd be 98 bottles of beer on the wall.

98 bottles of beer on the wall.
98 bottles of beer.
If one of those bottles should happen to fall,
there'd be 97 bottles of beer on the wall.
```

```
97 bottles of beer on the wall.
97 bottles of beer.
If one of those bottles should happen to fall,
there'd be 96 bottles of beer on the wall.
      .
      .
      .
1 bottles of beer on the wall.
1 bottles of beer.
If one of those bottles should happen to fall,
there'd be 0 bottles of beer on the wall.
```

A short name n is chosen for the do variable to make it easier to sing the program listing. The execution output shown is abbreviated after three full verses, with the last verse also given to show how the loop ends.

Testing if a Loop is Exited

The use of an exit statement makes clearer one example that previously seemed clearer with a go to statement. After executing a loop, it is sometimes needed to test if the loop went through all of its iterations or exited prior to that. This can be done nicely with a block construct and exit statement.

```
program exit_test

   implicit none
   real, dimension(5) :: x = &
      [ 1, 2, 3, 4, 5 ]

   find: block
      integer :: i
      do i = 1, 5
         if (x(i) > 2.2) then
            print *, "Location:", i
            exit find
         end if
      end do
      print *, "Not found"
   end block find

end program exit_test
```

2.5 Case Study: Numerical Integration I

The value of a definite integral is the area of a region of the plane bounded by the three straight lines. $x = a$, $y = 0$, $x = b$, and the curve $y = f(x)$ as shown in Figure 2-4. The better part of a semester in any calculus sequence is spent seeking analytic solutions to the area problem, that is, expressing the area by an algebraic or trigonometric expression. At the conclusion, the calculus student acquires a modest repertoire of useful functions that can be integrated in "closed form".

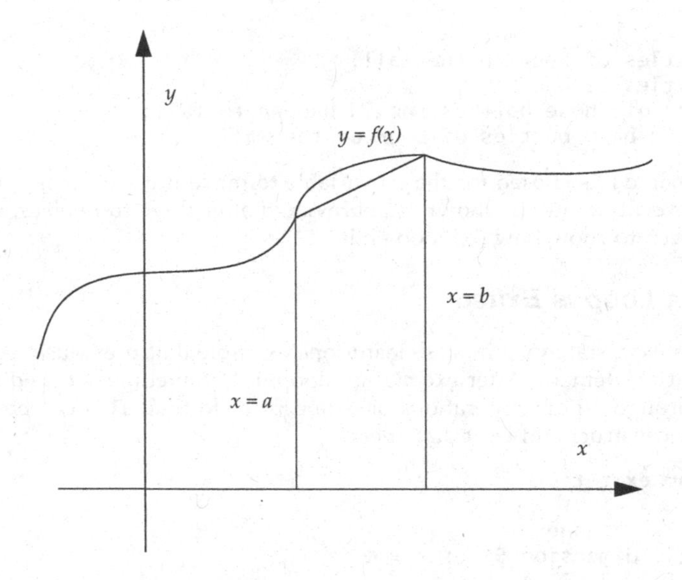

Figure 2-4 Trapezoidal approximation to the area under a curve

It turns out to be easier to approximate the area of such regions numerically, if you have a computer available. Moreover, the numerical approximation method works even for functions that cannot be integrated in closed form. If we replace the curve $y = f(x)$ by a straight line with endpoints a and b, the region in question is converted to a trapezoid, a simple four-sided figure whose area is given by the formula

$$A = (b - a) \times \frac{f(a) + f(b)}{2}$$

Of course, the area of this trapezoid is not exactly equal to the area of the original region with curved boundary, but the smaller the width of the trapezoid, the better the approximation.

Specifically, the problem we wish to solve is to find the area of one arch of the curve $y = \sin(x)$, that is, the area under this curve for x from 0 to π radians (180°) as shown in Figure 2-5. We will do it by writing a program to calculate trapezoidal approximations to the area, choosing a number of trapezoids sufficient to give the answer to three decimal places.

If we call the width of each trapezoid h, we have the relationship

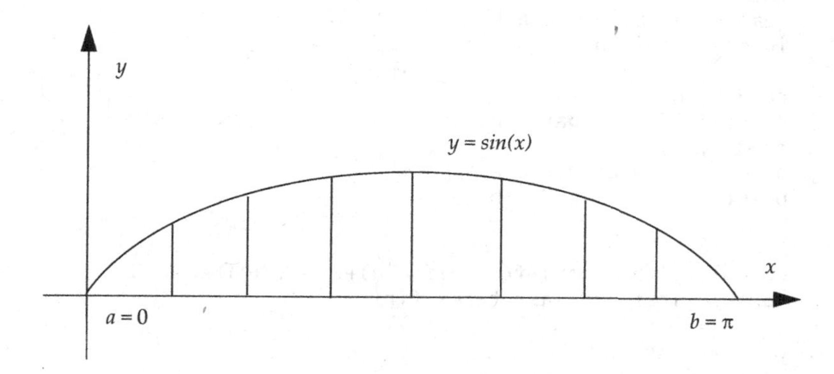

Figure 2-5 Approximating the area under the curve $y = sin(x)$

$$h = \frac{b-a}{n}$$

After a little algebra, the sum of the areas of the n trapezoids may be expressed by the formula

$$T_n = h\left(\frac{f(a)}{2} + f(a+h) + f(a+2h) + \cdots + f(b-h) + \frac{f(b)}{2}\right)$$

In the following program `integral`, the sum is formed by first computing

$$\left(\frac{f(a)}{2} + \frac{f(b)}{2}\right) = 0.5 \times [f(a) + f(b)]$$

Because a `do` variable must be an integer, we use the integer variable `i` that counts 1, 2, ..., $n-1$ and compute the expression `a + i*h` to obtain the sequence of values

$a + h, a + 2h, ..., a + (n - 1)h = b - h$

The program follows.

```
program integral
!  Calculates a trapezoidal approximation to an area
!  using n trapezoids.
!  n is read from the input file.

!  The region is bounded by lines x = a, y = 0, x = b,
!  and the curve y = sin(x).
!  a and b also are read from the input file.

   implicit none
```

```
   intrinsic :: sin
   real :: a, b, h, total
   integer :: i, n

   read *, n
   print *, "Input data  n:", n
   read *, a, b
   print *, "Input data  a:", a
   print *, "            b:", b

   h = (b - a) / n
!  Calculate the total f(a)/2+f(a+h)+...+f(b-h)+f(b)/2
!  Do the first and last terms first
   total = 0.5 * (sin(a) + sin(b))
   do i = 1, n - 1
      total = total + sin(a + i * h)
   end do

   print *, "Trapezoidal approximation to the area =", &
            h * total
end program integral
```

```
 Input data  n: 100
 Input data  a:    0.0000000E+00
             b:    3.1415901
 Trapezoidal approximation to the area =    1.9998353

 Input data  n: 1000
 Input data  a:    0.0000000E+00
             b:    3.1415901
 Trapezoidal approximation to the area =    1.9999995
```

Since these two answers differ by only one in the fourth decimal place, we may conclude that the approximation using 100 trapezoids is sufficiently accurate for our purposes, and that the approximation using 1000 trapezoids is accurate to more than four decimal places. (An alert reader may have noticed that the sixth and seventh decimal places in the echo of the input variable b are not the correct digits of π.) There is no need to rerun the program using more trapezoids to meet the limits of accuracy specified in the problem statement. The answer is 1.9999995 rounded to three decimal places to get 2.000. The input data for b was given using five decimal places as 3.14159 and the last two places echoed represent roundoff.

The `intrinsic` statement

The `intrinsic` statement consists of the keyword `intrinsic` followed by a double colon (`::`) and followed by a list of intrinsic procedure names. It can be used to indicate the use of any intrinsic procedure for documentation, but is usually not required.

2.6 Exercises

1. Hand simulate the execution of the following statements, keeping track of the value of n and prod after the execution of each statement.

```
integer :: n, prod
prod = 1
do n = 2, 4
   prod = prod * n
end do
```

2. What output is produced by the following program?

```
program exercise
   implicit none
   intrinsic :: modulo
   integer :: m
   do m = 1, 20
      if (modulo(m, 2) /= 0) then
         print *, m
      end if
   end do
end program exercise
```

3. What is the value of the variable total at the conclusion of each of the following loops?

```
integer :: n, total

total = 0
do n = 1, 10
   total = total + 1
end do

total = 0
do n = 1, 5
   total = total + n * n
end do

total = 0
do n = 1, 14, 2
   total = total + n * n
end do

total = 0
do n = 5, 1, -1
   total = total + n
end do
```

4. An integer is a perfect square if it is the square of another integer. For example, 25 is a perfect square because it is 5 × 5. Write a program to selectively print those numbers less than 100 that are *not* perfect squares. Sample output for this program should look like the following.

```
2
3
5
6
7
8
10
.
.
.
99
```

5. Read integers from the input file until the value zero is read. Then print the number in the file just before the first zero value. Sample input data might be

```
3
7
2
10
0
9
4
0
5
```

Sample output for this input data is

```
Input data:  buffer  3
Input data:  buffer  7
Input data:  buffer  2
Input data:  buffer  10
Input data:  buffer  0
The last number before the first zero is 10
```

6. Write a program that prints the smallest power of 3 that exceeds 5000.

7. In 1970, the population of New Jersey was 7,168,192 and it was increasing at the rate of 18% per decade. The area of New Jersey is 7521 square miles. On the basis of the 18% growth rate continuing indefinitely into the future, predict the population of New Jersey every decade from 1980 on. Stop the predictions when the average number of square feet per person is less than 100. Print out all estimates. Execution of the program should produce something like the following.

```
year     population     sq ft / person
1980       8458466.            24789.6
1990       9980990.            21007.3
```

. . .
. . .
. . .

For partial confirmation of the validity of the prediction model, look up the 1980, 1990, and 2000 census data for New Jersey and compare the actual data with your program's predictions.

8. The mathematical expression

$$\left(1 + \frac{1}{n}\right)^n$$

produces better and better approximations to the famous mathematical constant e = 2.718281828459045... as n gets large. However, the computed result of this expression may be disappointingly inaccurate if the selected real kind does not permit many significant digits of $1/n$ to be retained in the sum $1+1/n$. Write a program to calculate the expression

$$\left(1 + \frac{1}{n}\right)^n$$

for n taking on successive powers of two: 1, 2, 4, 8, 16, ... and successive powers of three: 1, 3, 9, 27, 81, ... Run this program using each of the real kinds available on your computer.

2.7 The stop Statement

The **stop statement** causes execution of a program to stop. With the use of modern control constructs, a program usually should stop by coming to the end of the program. However, there are some occasions where the stop statement is very convenient to use. For example, when print statements are inserted for debugging, it is often desirable to stop the program after a few such statements are executed or after the first few iterations of a loop are executed. Also, when severe errors are detected in the middle of a procedure that is being executed, it is much easier to execute a stop statement than exit out through what may be many layers of nested subroutine calls or function references.

```
read *, income
if (income < 0) then
   print *, "Error: income is less than zero."
   stop
end if
```

2.8 The go to and continue Statements

The if, case, and do constructs are sufficient to build almost any program. However, there are some rare occasions that seem to require a direct branch to a different part of the program. This can be done with the go to statement, which can transfer control to a continue statement.

The form of the go to statement is

go to *label*

where *label* is a string of one to five decimal digits, at least one of which must be nonzero. When the statement is executed, control is passed to the continue statement with that label. The continue statement with the label should be after the go to statement; branching backward produces code that is very difficult to understand.

The continue statement has the form

label continue

Branching from outside a construct (if, case, do, block, associate, or where) to a continue statement inside the construct is not permitted.

One situation in which it might be convenient to use a go to statement is when a serious error condition occurs inside a fairly complex construct, such as nested do loops.

3 Modules and Procedures

Large programs are extremely difficult to debug and maintain unless they are split into independent parts. Even relatively short programs are greatly improved when their component parts are refined as procedures. Modules provide a place to put data declarations so that they can be used and shared by programs. Modules also provide the place to put a Fortran **procedure**, which is either a function or a subroutine, and to put definitions of user-defined types; these are basic building blocks of a program and are usually used by more than one part of a program.

Modules are especially useful when building a "package" or "library" of data and procedures that may be accessible to many different programs.

Submodules are discussed in 8.1.

3.1 Modules

A **module** is a program unit that is not executed directly, but contains data specifications and procedures that may be utilized by other program units via the `use` statement.

The general form of a module is:

```
module name
  use statements
  declaration statements
contains
  subroutines and functions
end module name
```

Style note: The `implicit none` and the `private` statements should appear in every module unless the module consists of only `use` statements; such a module is used to collect the information in several other modules.

Writing and Using Modules

To begin with a very simple example, one use of a module is to include the definition of constants that might be useful in programs. The module `math_module` contains the values of π, e, and γ; of course, it could contain many more useful constants. Note that these constants have default kind, regardless of the number of decimal digits that appear.

```
module math_module

   implicit none
   private
   real, public, parameter :: pi = &
         3.141592653589793238462643383279502884 1972
   real, public, parameter :: e = &
         2.718281828459045235360287471352662497757 2
   real, public, parameter :: gamma = &
         0.577215664901532860606512090082402431042 2

end module math_module
```

Any program that needs these constants can simply use the module.

```
program circle

   use math_module
   implicit none
   real :: radius = 2.2, area

   area = pi * radius ** 2
   print *, area

end program circle
```

It is also possible to declare variables in a module. The module `declarations_module` declares logical variables `flag_1` and `flag_2`, which could then be used in any program that uses the module.

```
module declarations_module

   implicit none
   private
   logical, public :: flag_1, flag_2

end module declarations_module

program using_modules

   use declarations_module
   implicit none

   logical, parameter :: f = .false.
   flag_1 = f
   flag_2 = .not. flag_1
      . . .
```

```
end program using_modules
```

Most implementations require that a module be compiled before any program that uses the module is compiled. Thus, if the module and program above are placed in the same source file, the module must come before the program. Also, the `use` statement is required, even if both module and program are in the same file.

Private and Public Access

Everything declared in a module has **access**, which is either private or public. Private things are known only within the module and public things can be accessed by any program or procedure that uses the module.

Each parameter, variable, and type (6.2) in a module should have either the `public` or `private` attribute in its declaration. It is possible for the programmer of a module to use the access statements to restrict the variables and procedures in the module that are accessible outside the module. This is done to "hide" implementation details in the module and is accomplished by declaring things private. A nice example of this is the module for computing with big integers discussed in 9.6. In the simple examples shown so far, everything is declared public.

The `private` Statement

The **`private` statement** in a module indicates that all things not declared public are private; without it, the default is that all things are public.

> *Style note:* A private module should have a `private` statement to prevent things in the module from being passed on.

The `protected` Attribute

If a variable in a module is declared to be public, its access can be partially restricted by further giving it the `protected` attribute in the declaration. This means that the value of the variable is accessible but cannot be changed. For example, a module might have a logical variable that indicates whether certain values in the module have been initialized or not. It would be acceptable to have the value of the variable accessible outside the module, but it should not be changed outside the module.

```
logical, public, protected :: &
      variables_have_been_initialized = .false.
```

The `use` Statement

The simple form of the **`use` statement** is just the keyword `use` followed by a module to be used, as illustrated by a previous example. A `use` statement may appear in a program, subroutine, function, another module, or a procedure interface.

However, with the use statement, there are two ways to affect the way that names in a module are accessed by another program unit. The first is that the names used in the module may be changed in the program unit using the module. This may be necessary because the program is using two or more modules that contain declarations of the same name. Or it simply may be desirable to change the name to suit the taste or needs of the programmer of the program unit.

For example, in a subroutine using module math_module, the programmer may decide that the name e is too short to allow a clear understanding of its purpose. This can be fixed by renaming the variable e to the longer name logarithm_base with the use statement.

```
use math_module, logarithm_base => e
```

Any number of rename clauses may appear in the use statement.

The second way to affect the names accessed in a module is to have an only clause in the use statement. In the program circle, only the constant π is needed. It is possible to prevent other names in the module from conflicting with names in the program; this can be accomplished with the use statement.

```
use math_module, only : pi
```

If, in addition, it were desirable to use and rename the parameter e to logarithm_base, this could be done with the statement:

```
use math_module, only : pi, logarithm_base => e
```

There can be many names, with or without renaming, in a list after the colon. A use statement can refer to only one module, but there can be more than one use statement in a program for a module.

```
use m, only : x
use m, only : y
```

3.2 Procedures

There are two kinds of **procedures**: functions and subroutines. A **function** looks much like a Fortran program, except that it begins with the keyword function instead of the keyword program. Once written, a function is used just like the built-in functions discussed in 1.5 to compute a value that may be used in any expression. A **subroutine** also looks like a program or a function, except that the first line begins with the keyword subroutine. A subroutine may be used to perform any computation and is invoked by executing a call statement.

Style note: All procedures should be placed in a module or after the contains statement in a main program or module procedure.

Style note: Self-contained subtasks should be written as procedures.

Functions and subroutines whose first statements contain the keyword `recursive` are permitted to call themselves directly or indirectly; recursion (3.16) is used to write clear and simple programs for what might otherwise be difficult programming tasks.

Functions and subroutines whose first statements contain the keyword `elemental` allow the programmer to more simply write a procedure that handles an array of values on an element-by-element basis (8.7).

The keyword `pure` (3.7) on a function or subroutine statement indicates that the procedure has no side effects.

Style note: Most functions should be pure, with or without the keyword.

An **access statement** consists of either `private` or `public` followed by a colon and a list of the names of procedures in the module.

3.3 Subroutines

Suppose the task at hand is to read in three real numbers and print them in ascending order. The main steps needed to accomplish this task are (1) read in the numbers, (2) sort them, and (3) print them. The program `sort_3` does this.

```
program sort_3

   implicit none
   call read_the_numbers()
   call sort_the_numbers()
   call print_the_numbers()

end program sort_3
```

It seems obvious (because the names are chosen well) that it performs the three steps described above needed to solve the problem. However, if you try to compile this into an executable program, you will be told that the three procedures are missing. We must provide statements that directly reflect these three steps and put the details elsewhere.

The `call` Statement

The **`call` statement** is used to indicate that the computation represented by a subroutine is to be performed. The keyword `call` is followed by the name of the subroutine and by a list of arguments (3.5) in parentheses. The program `sort_3` contains three `call` statements.

Writing a Subroutine

A subroutine is very similar to a program except that the first statement is a **subroutine statement** that begins with the keyword subroutine and ends with a list of arguments in parentheses. The parentheses should appear even if there are no arguments. The last statement of a subroutine is the **end subroutine statement**, which contains the name of the subroutine.

The subroutine read_the_numbers consists of the subroutine statement, the statements that read and echo the numbers, and the end subroutine statement that terminates the subroutine.

```
subroutine read_the_numbers()
   read *, n1, n2, n3
   print *, "Input data  n1:", n1
   print *, "            n2:", n2
   print *, "            n3:", n3
end subroutine read_the_numbers
```

It is not necessary to declare the variables n1, n2, and n3 because they will be declared elsewhere.

The other two subroutines are constructed similarly (see below).

3.4 Putting Procedures in a Module

We now show how the program to sort three numbers can be organized using a module to contain the subroutines. The module also will contain the declaration of the four private variables n1, n2, n3, and temp because they are used by the procedures in the module. The module sort_3_module also contains the three subroutines after a contains statement.

Procedures appear just before the last end statement of the module containing them and they are preceded by a **contains statement**, which consists of simply the keyword contains.

```
module sort_3_module

   implicit none
   private
   real :: n1, n2, n3
   real :: temp

   public :: read_the_numbers,  &
             sort_the_numbers,  &
             print_the_numbers

contains
```

```
subroutine read_the_numbers()
   read *, n1, n2, n3
   print *, "Input data  n1:", n1
   print *, "            n2:", n2
   print *, "            n3:", n3
end subroutine read_the_numbers

subroutine sort_the_numbers()
   if (n1 > n2) then
      temp = n1
      n1 = n2
      n2 = temp
   end if
   if (n1 > n3) then
      temp = n1
      n1 = n3
      n3 = temp
   end if
   if (n2 > n3) then
      temp = n2
      n2 = n3
      n3 = temp
   end if
end subroutine sort_the_numbers

subroutine print_the_numbers()
   print *, "The numbers, in ascending order, are:"
   print *, n1, n2, n3
end subroutine print_the_numbers

end module sort_3_module
```

The following program uses the module to sort three numbers. The statement

```
use sort_3_module
```

indicates that there are procedures or data in a module called `sort_3_module` that are needed by the program. Indeed, the computations will each be done with subroutines that are in the module.

```
program sort_3

   use sort_3_module
   implicit none

   call read_the_numbers()
   call sort_the_numbers()
   call print_the_numbers()
```

```
end program sort_3
```

Running the program produces

```
Input data  n1:    2.2000000
            n2:    7.6999998
            n3:    5.5000000
The numbers, in ascending order, are:
  2.2000000   5.5000000   7.6999998
```

3.5 Arguments

Something worth noticing is that there are three lines that occur three times in the subroutine sort_the_numbers, all doing the same kind of operation, namely, swapping the values of two variables if they are in the wrong order. This illustrates the second good reason to use a procedure: to write some statements once and use them many times, either within the same program or in different programs. In this case, the computation that is performed three times is represented the first time by the three statements:

```
temp = n1
n1 = n2
n2 = temp
```

However, each time this swapping operation occurs in the subroutine, different named variables are involved. This is no obstacle if a subroutine with arguments is used as illustrated by the subroutine named swap.

```
subroutine swap(a, b)
   real, intent(in out) :: a, b
   real :: temp
   temp = a
   a = b
   b = temp
end subroutine swap
```

To call this subroutine, values are sent to it by placing them in parentheses after the name of the subroutine in the call statement. Thus, to swap the values of n1 and n2, use the statement

```
call swap(n1, n2)
```

n1 and n2 are called **arguments**. Argument passing applies to both subroutines and functions and so is described in more detail in 3.8.

The subroutine sort_the_numbers can now use swap.

```
subroutine sort_the_numbers()
```

```
      if (n1 > n2) then
         call swap(n1, n2)
      end if
      if (n1 > n3) then
         call swap(n1, n3)
      end if
      if (n2 > n3) then
         call swap(n2, n3)
      end if
   end subroutine sort_the_numbers
```

The subroutine `swap` must be made available either by placing it in the same module with `sort_the_numbers`, in which case it can be declared `private` because it is used only within the same module, or placing it in a separate module and declaring it `public`.

Dummy Arguments and Local Variables

There are two new variables `a` and `b` in the subroutine `swap` that serve as place holders for the two numbers to be swapped. These are **dummy arguments** and must be declared in the subroutine even if they have the same name as a variable declared in the containing module.

The variable `temp` is used only in the subroutine `swap`. By declaring `temp` to be type real within the subroutine `swap`, we make this variable **local** to the subroutine, so that its value will not be confused with any value outside the subroutine. The declaration of `temp` can be removed from the subroutine `sort_the_numbers`.

Argument Intent

In Fortran you should indicate the **intent** of use of each dummy argument of a subroutine or function unless it is a dummy procedure. The intent may be `in`, which means that the dummy argument cannot be changed within the procedure; it may be `out`, which means that the actual argument must not be used until given a value in the procedure and is used to pass a value back to the calling program; or it may be `in out`, which means that the dummy argument is expected both to receive an initial value from and return a value to the corresponding actual argument. Thus, for dummy arguments with intent `out` or `in out`, the corresponding actual argument must be a variable.

The intent of a pointer (10.1) dummy argument applies to the pointer itself, not the target. That is, the value (of the target) of a dummy pointer with intent `in` may be changed within the procedure, but the pointer status may not.

The intent is an attribute given to an argument when it is declared within the procedure.

The intent attribute is provided to make the program more easily understood by a human reader and to allow the compiler to catch errors when the programmer violates the stated intent.

Style note: All dummy arguments (except procedures) should have their intent declared.

Style note: All dummy arguments (except procedures) to a function should have intent `in` to help enforce the style of writing only pure procedures.

The `value` Attribute

If a dummy argument has the `value` attribute, it may be changed, but this will not effect the corresponding actual argument. This is similar to declaring `intent`, but is a separate attribute. It is a useful attribute when describing C functions (8.10), because this is the way most C argument passing works unless pointers are used.

```
program value_test

   implicit none
   real :: x = 1.1
   call s(x)
   print *, x  ! produces 1.1

contains

subroutine s(d)
   real, value :: d
   d = 2*d
   print *, d  ! produces 2.2
end subroutine

end program value_test
```

Exercises

1. Write a module named `swap_module` that contains only the subroutine `swap`.
2. Remove `swap` from the module `sort_3_module` and rename it `sort_module`.
3. Write and test a public module `sort_3_module` that uses `swap_module` and `sort_module`.
4. Write a subroutine `sort_4_numbers` that arranges the four integer variables `i1`, `i2`, `i3`, and `i4` into ascending order. Test the subroutine by putting it in a program that reads four numbers, calls the subroutine, and prints the sorted values.
5. Write a subroutine that reads in values for a loan principal amount p, an annual interest rate r_{annual}, and the number of months m in which the loan is to be paid off. The monthly payment is given by the formula

$$pay = \frac{r \times p(1+r)^m}{(1+r)^m - 1}$$

where the monthly interest rate $r = r_{annual}/12$. The subroutine should print out a monthly schedule of the interest, principal paid, and remaining balance. Test the subroutine with a program that calls it with p = \$106,500, r_{annual} = 7.25%, and m = 240 months.

3.6 Functions

If the purpose of a procedure is to compute one value (which may be a compound value consisting of a whole array or structure) and the procedure has no side effects (3.7), a function is the sort of procedure to use. The value of a function is computed when the name of the function, together with its arguments, is placed anywhere in an expression.

To illustrate a simple use of a function, suppose the task is to print out a table of values of the function

$$f(x) = \left(1 + \frac{1}{x}\right)^x$$

for values of x equal to 1, 10, 100, ..., 10^{10}. A program to do this is

```
module f_module

   implicit none
   private
   integer, parameter, public :: largest_power = 10
   public :: f

contains

function f(x) result(f_result)

   real, intent(in) :: x
   real :: f_result

   integer, parameter :: kind_needed = &
      selected_real_kind(largest_power + 1)

   f_result = (1 + 1 / real(x, kind_needed)) ** x

end function f
```

```
end module f_module

program function_values

   use f_module
   implicit none
   real :: x
   integer :: i

   do i = 0, largest_power
      x = 10.0 ** i
      print "(f15.1, f15.5)", x, f(x)
   end do

end program function_values
```

```
            1.0        2.00000
           10.0        2.59374
          100.0        2.70481
         1000.0        2.71692
        10000.0        2.71815
       100000.0        2.71827
      1000000.0        2.71828
     10000000.0        2.71828
    100000000.0        2.71828
   1000000000.0        2.71828
  10000000000.0        2.71828
```

In this program the evaluation of the function occurs once for each execution of the do construct, but the expression that evaluates the function occurs only once. In this case, a function is used to put the details of evaluating the function in another place, making the program a little easier to read. When this is done, there is also the advantage that if a similar table of values is needed, but for a different function, the main program does not need to be changed; only the function needs to be changed.

This function illustrates an interesting use of the `selected_real_kind` intrinsic function. In the function, the intermediate result $1 + 1/x$ must be computed to get the desired answers. For $x = 10^{10}$, this value is 1.0000000001, which has 11 significant digits, so a kind of real must be used that will hold this many digits. If a real kind with fewer significant digits is used, the expression `1 + 1/x` may evaluate as 1.00000, yielding an incorrect value for `f_result`. Both the largest power of x used and the kind needed to compute the function for this largest power are provided as parameters (named constants).

The type conversion

```
real(x, kind_needed)
```

converts the already real value `x` to a real with kind of the required precision, and the rules for mixed mode arithmetic guarantee that at least this precision is used throughout the calculation. `kind_needed` must be a literal or named constant.

Style note: Whenever an integer is used as a kind number, it should be a parameter, not a literal constant. It must not be a variable.

Writing a Function

A function is almost like a subroutine except that its first statement uses the keyword `function`. Like a subroutine, it may have arguments that are written in parentheses in the **`function` statement**. This is followed by the keyword `result` and the name of the result variable. The last statement of a function is the **`end function` statement**, which contains the name of the function.

A difference between a subroutine and a function is that a function must provide a value that is returned as the value of the function. This is done by assigning a value to a **result variable** during execution of the function. This result variable is indicated by placing its name in parentheses at the end of the `function` statement following the keyword `result`. The result variable is declared within the function and is used just like any other local variable, but the value of this variable is the one that is returned as the value of the function to the program using the function. Intent is not declared for the result variable—its appearance in the result clause effectively makes its intent `out`. The function `f` computes the values required in our example and uses the result variable `f_result` to hold the result.

Invoking a Function

A programmer-defined function is called by writing its name, followed by its arguments, in any expression in the same manner that a built-in function is invoked.

Exercises

1. Write a function `median_of_3` that selects the median of its three integer arguments. If all three numbers are different, the median is the number that is neither the smallest nor the largest. If two or more of the three numbers are equal, the median is one of the equal numbers.

2. Write a function `average_of_4` that computes the average of four real numbers.

3. Write a function `cone_volume(r, h)` that returns the volume of a cone. The formula for the volume of a cone is $V = \pi r^2 h/3$, where r is the radius of the base and h is its height.

4. Write a function `round(x, n)` whose value is the real value `x` rounded to the nearest multiple of 10^n. For example, `round (463.2783, -2)` should be 463.28, which has been rounded to the nearest hundredth.

3.7 Pure Procedures and Side Effects

When a procedure is executed, a **side effect** is a change in the status of the program that is something other than just computing a value to return to the calling procedure. Examples are changing a variable declared in a program or module above the `contains` statement or reading data from a file.

The programmer may indicate that a procedure has no side effects by putting the keyword `pure` in the function or subroutine statement. All elemental procedures (8.7) must be pure, unless the keyword `impure` also appears.

Style note: All functions should be pure. A reasonable exception is that a function might contain `print` statements for debugging.

When the following Fortran language rules are followed, most side effects will not occur. Some of the rules involve features described later.

1. All dummy arguments in a function (except procedures, which never have an intent attribute) must have intent `in`.
2. A local variable must not have the `save` attribute or be initialized by a declaration statement.
3. Any subroutine that is called, including a defined assignment (9.1) must be pure.
4. There is no input/output statement, except for an internal `read` or `write` statement (11.3).
5. The `use` statement permits a function to import names from a module without placing them in the dummy argument list. The following additional rules are necessary to prevent side effects with such variables. Any variable that is accessed from a module by a `use` statement or has intent `in` must not appear as any of the following:
   a. the variable on the left of an assignment statement
   b. an input item in an internal `read` statement
   c. a character string used as the file in an internal `write` statement
   d. the variable assigned a value as an `iostat` or `iomsg` specifier in an input/output statement using an internal file
   e. either the pointer or the target in a pointer assignment statement
   f. the right side of an assignment statement, if the left side is of derived type with a pointer component
   g. the object to be allocated or deallocated or the status variable in either an `allocate` or `deallocate` statement

h. an actual argument in a procedure reference where the corresponding dummy argument is intent `out` or `in out`.

A procedure that is invoked in any of the following circumstances must be pure; that is, the procedure heading must contain the keyword `pure` or the keyword `elemental`.

1. a function referenced in a specification statement
2. a procedure that is passed as an actual argument to a pure procedure
3. a procedure referenced in a pure procedure, including those referenced by any function, a defined operator (9.2), or defined assignment (9.1).

3.8 Argument Passing

One of the important properties of both functions and subroutines is that information may be passed to the procedure when it is called and information may be returned from the procedure to the calling program when the procedure execution ends. This information passing is accomplished with procedure **arguments** and, in the case of a function, the function result. A correspondence is set up between **actual arguments** in the calling program and **dummy arguments** in the procedure. The corresponding arguments need not have the same name, and the correspondence is temporary, lasting only for the duration of the procedure call.

Agreement of Arguments

In this subsection, we try to emphasize general principles, but for the sake of having all the important rules in one place, we list exceptions needed to implement these language features along with forward references to the sections where they are discussed.

Except for dummy arguments declared as optional (see below), the number of actual and dummy arguments must be the same. Each actual argument corresponds to a dummy argument. The default correspondence is the first actual argument with the first dummy argument, the second with the second, etc. However, keyword-identified arguments (see below) can be used to override the default, and provide clear, order-independent specification of the correspondence between actual and dummy arguments.

The data type and kind parameter of each actual argument must match that of the corresponding dummy argument.

Additionally, if the dummy argument is a pointer (10.1), the actual argument must be a pointer.

If the subroutine or function is generic (8.6), there must be exactly one specific procedure with that generic name for which all the above rules of agreement of actual and dummy arguments are satisfied (however, keyword actual arguments also can be used to determine which procedure is specified). For given actual arguments, Fortran selects that specific procedure for which there is agreement of actual and dummy arguments.

Passing Arguments to Dummy Arguments with Intent `out`

If an actual argument is passed to a dummy argument that has intent `out` or intent `in out`, it must be a variable (which includes an array name, an array element, an array substring, a structure component, or a substring) so that it makes sense to give it a value. Any reference to the corresponding dummy argument in the subroutine causes the computer to behave as if the reference were to the corresponding actual argument supplied by the calling program, unless the dummy argument has the `value` attribute. Statements in the subroutine causing changes to such a dummy argument cause the same changes to the corresponding actual argument.

> *Style note:* A dummy argument in a function should not be intent `out` or intent `in out`.

> *Style note:* A dummy argument that is intent out should be given a value during execution of the procedure.

Passing Arguments to Dummy Arguments with Intent `in`

An actual argument that is a constant (either literal or named) or an expression more complicated than a variable must correspond to a dummy argument with intent `in`. The dummy argument then must not have its value changed during execution of the procedure. There is no way to pass a value back to the calling program using such an argument.

An Example of Passing Variables

Let us look again at the subroutine `swap` discussed earlier and how it is used in the program `sort_3` in 3.4. In the `subroutine` statement, the subroutine name `swap` is followed by a list `(a, b)` of variables enclosed in parentheses. The variables `a` and `b` in that list are the dummy arguments for the subroutine `swap`. They have intent `in out`.

Suppose that in executing the first `read` statement of the subroutine `read_the_numbers`, the computer reads and assigns to the variable `n1` the value 3.14, assigns to the variable `n2` the value 2.718, and assigns to the variable `n3` the value 1.414. Since 3.14, the value of `n1`, is greater than 2.718, the value of `n2`, the computer executes the call statement

```
call swap(n1, n2)
```

The effect of this `call` statement is as if it were replaced by the following statements.

```
! Copy-in phase
a = n1
b = n2
```

```
temp = a
a = b
b = temp

! Copy-out phase
n1 = a
n2 = b
```

In this example, the dummy arguments a and b both have intent in out. For a procedure with arguments with intent out, the copy-in phase may be skipped and for a procedure with arguments with intent in, the copy-out phase may be skipped.

An Example of Passing Expressions

Suppose a function is to be written that computes the following sum of certain terms of an arithmetic progression:

$$\sum_{i=m}^{n} (s + d \times i)$$

The arguments to this function are *m, n, s* (the starting value), and *d,* the difference between terms. A function to do this computation is contained in the program series and a module is not used in this example in order to keep it a little simpler.

```
program series

   implicit none
   integer, parameter :: n = 100
   print *, series_sum(n+300, 2*n+500, 100.0, 0.1)

contains

function series_sum(m, n, s, d)  result(series_sum_result)

   integer, intent(in) :: m, n
   real, intent(in) :: s, d
   real :: series_sum_result
   integer :: i

   series_sum_result = 0
   do i = m, n
      series_sum_result = series_sum_result + s + i * d
   end do

end function series_sum

end program series
```

which produces the answer 46655.0. All four actual arguments in the call of `series_sum` are constants and therefore may be passed to the intent `in` arguments of the function `series_sum`.

Keyword Arguments

With the use of **keyword arguments**, it is not necessary to put the arguments in the correct order, but it is necessary to know the names of the dummy arguments. The same computation may be made using the statement

```
print *, series_sum(d=0.1, m=400, n=700, s=100.0)
```

It is even possible to call the function using keywords for some arguments and not for others. In this case, the rule is that all actual arguments prior to the first keyword argument must match the corresponding dummy argument correctly and once a keyword argument is used, the remaining arguments must use keywords. Thus, the following is legal:

```
print *, series_sum(400, 700, d=0.1, s=100.0)
```

Optional Arguments

In our example computation of an arithmetic series, a common occurrence would be that the value of `m` is 0. It is possible to indicate that certain arguments to a procedure are **optional arguments** in the sense that they do not have to be present when the procedure is called. An optional argument must be declared to be such within the procedure; usually, there would be some statements within the procedure to test the presence of the optional argument on a particular call and perhaps do something different if it is not there. In our example, if the function `series_sum` is called without the argument `m`, the value zero is used. To do this, the intrinsic function `present` is used to test whether an argument has been supplied for the dummy argument `m`, and if an actual argument is not present, the lower bound for the sum is set to zero. To handle both cases with the same `do` loop, a different variable, `temp_m`, is used to hold the lower bound. One reason a different variable is used is that *a dummy argument corresponding to an actual argument that is not present must not be given a value within the procedure.* The other reason is that all function arguments are intent `in`, and so cannot be changed anyway.

If the actual argument corresponding to an optional dummy argument is a null pointer (10.1) or an unallocated allocatable array (4.1), the argument is considered to be not present.

```
function series_sum(m, n, s, d)  &
      result(series_sum_result)

   integer, optional, intent(in) :: m
   integer, intent(in) :: n
   real, intent(in) :: s, d
```

```
      real :: series_sum_result
      integer :: i, temp_m

      if (present(m)) then
         temp_m = m
      else
         temp_m = 0
      end if

      series_sum_result = 0
      do i = temp_m, n
         series_sum_result = series_sum_result + s + i * d
      end do

   end function series_sum
```

This new version of the function can now be called with any of the following statements, all of which compute the same sum:

```
   print *, series_sum(0, 700, 100.0, 0.1)
   print *, series_sum(0, 700, d=0.1, s=100.0)
   print *, series_sum(n=700, d=0.1, s=100.0)
   print *, series_sum(d=0.1, s=100.0, n=700)
   print *, series_sum(m=0, n=700, d=0.1, s=100.0)
```

Procedures as Arguments

An actual argument and the corresponding dummy argument may be a procedure. The actual argument itself may be a dummy procedure. Only a few intrinsic procedures may be passed as actual arguments. The actual argument may be internal, but usually is in a module.

3.9 Interface Blocks

In a function or subroutine that has a procedure as a dummy argument, the dummy argument must be "declared", much as every other dummy argument is declared. However, to "declare" a procedure, quite a bit of information should be provided. An **interface block** is used for this purpose. The purpose of the interface block is to provide the information necessary to a calling program to tell whether the call is correct. An interface block basically consists of the procedure itself with all of the executable code and declarations of local variables removed, leaving all of the information about its arguments and the result returned if it is a function.

Interface blocks also are used to describe C functions that are called from a Fortran program (8.10), external procedures (not discussed in this book), procedures that extend Fortran (9), derived-type procedure components (12.3), and type-bound procedures (12.3).

In 3.14, the numerical integration routine has a dummy argument that is the function to be integrated. In this case, the function has one real argument and the result is real. Thus, the interface block for this dummy argument is

```
interface
function f(x) result(f_result)
   real, intent(in) :: x
   real :: f_result
end function f
end interface
```

In this case, the interface block contains almost the whole function because there is only one executable statement in the function. In general, of course, the executable part of a function may be fairly lengthy and the interface block will be much smaller than the whole function.

An interface block has its own scope, so it is sometimes necessary to use the `import` statement (3.11) to provide access to things needed in the interface, such as a parameter.

3.10 Exercises

1. Write a program that tests `cone_volume` (Exercise 3 of 3.6) using keywords to call the function with arguments in an inverted order.

2. Rewrite the function `cone_volume` (Exercise 3 of 3.6) to make the radius an optional argument with a default value of 1 if it is not present. Test the revised function by using it both with the argument present and with the argument missing.

3.11 Using a Function in a Declaration Statement

Intrinsic functions are allowed in declarations—for example, in the specification of the size of an array. Also, in some circumstances, it is possible to invoke a user-defined function. A simple example is to create a parameter equal to the logarithm (base 10) of the parameter `size`.

```
real, parameter :: log_10_size = log10(size).
```

3.12 The return Statement

The **return statement** causes execution of a procedure to terminate with control given back to the calling procedure. With the use of modern control constructs, a procedure usually should terminate by coming to the end of the procedure. However, there are situations in which it is better to use a return statement than introduce a complicated set of nested if constructs. Most of the programs in this book are too simple to require use of the return statement.

3.13 Scope

The **scope** of a name is the set of lines in a Fortran program where that name may be used and refer to the same parameter, variable, procedure, or type. In general, the scope of a parameter or variable declared in a program or module above the contains statement extends throughout that program from the program or module statement to the corresponding end statement, including any contained procedures, except those in which the name is used to declare some other object in the procedure.

A name declared in a procedure has scope extending only from the beginning to the end of that procedure, not to any other procedure (procedures contained in module procedures, not discussed in this book, are an exception).

A name declared in a block construct (2.1) is known only within the construct.

Names declared with the public attribute above the contains statement in a module have larger scope. This scope includes all modules and programs that use the module and do not exclude the name with an only clause. These ideas are illustrated by the following module segment.

```
module m
   implicit none
   public :: s
   integer, private :: a, b
   . . .
contains

subroutine s()
   real :: b
   . . .
   print *, a, b
   . . .
```

The values of a and b are printed in the subroutine. a is the integer variable declared in the module; its scope includes the subroutine because it is not redeclared. However, it is a real value that is printed for b, which is the b declared in the subroutine s. The scope of the integer b declared in the module does not include the subroutine s. That is, there are two variables with the name b, an integer variable b, whose scope is the module and does not include the subroutine s, and a real variable b, whose scope consists of the subroutine s only.

The name of a procedure, its number and type of arguments, their names for use only in keyword actual arguments, as well as the type of its result variable if it is a function, are considered as declared in the containing module or program, and its scope extends throughout the module or program. Therefore, a procedure can be called by any procedure in the module or program and, if it is public in a module, any procedure in a program or another procedure that uses the module.

The save Attribute

Unless something special is done by the programmer, the value of a variable that is local to a procedure is not saved between calls to the procedure. Suppose it is desirable to have a variable in a subroutine that counts the number of times the subroutine is called; this might be useful for debugging, for example.

```
subroutine s()
   integer, save :: call_count = 0

   call_count = call_count + 1
   print *, "This is execution #", call_count,  &
            "of subroutine s."
!        . . .
end subroutine s
```

In this case, the value of the local variable call_count is saved between calls of the subroutine because it is declared with the save attribute.

If a variable is given an initial value in a subroutine or function, it has the save attribute whether or not the keyword save appears.

All variables declared in a module have the save attribute.

The import Statement

The import statement is used to gain access to an entity that is not in the current scope. This is frequently used inside an interface block (3.9) since an interface block has its own scope. For example, supposed a kind number is declared in a module and is needed to declare a dummy argument in an interface block. The import statement in the following code makes it accessible inside the interface block. A use statement inside the interface block also could be used to achieve the same purpose.

```
module param_mod
   implicit none
   integer, parameter :: kk = kind(0.0)
end module param_mod

program import_prog
   use param_mod
   implicit none
```

```
   interface
      subroutine s(x)
      import :: kk
      real(kind=kk), intent(in) :: x
      end subroutine
   end interface
   . . .
end program import_prog
```

3.14 Case Study: Numerical Integration II

In 2.5, we wrote a program `integral` to approximate the definite integral

$$\int_a^b f(x)dx$$

by dividing the interval from a to b into n equal pieces, approximating the curve with straight lines, and computing the sum of the areas of the n trapezoids with the formula

$$T_n = h\left(\frac{f(a)}{2} + f(a+h) + f(a+2h) + \cdots + f(b-h) + \frac{f(b)}{2}\right)$$

In the program `integral`, the values for a, b, and n were read as input data. Now that we have procedures, a better approach is to write a function `integral` with arguments `a`, `b`, and `n`. The other problem with the program `integral` is that the name of the function to be integrated (`sin`, in the example), was "hard-wired" into the source code and could not be changed without rewriting and recompiling the program. Since it is possible to pass a procedure as an argument, we can make the name of the function an additional argument `f` to our function integral. The executable statements of the function `integral` use the dummy function argument `f` in place of the particular function `sin`, resulting in the following program `integrate`.

```
module integrate_module

   implicit none
   private
   public :: integral

contains

function integral(f, a, b, n)  result(integral_result)
!  Calculates a trapezoidal approximation to an area
!  using n trapezoids.

!  The region is bounded by lines x = a, y = 0, x = b,
```

```
!  and the curve y = f(x).

   interface
      function f(x) result(f_result)
         real, intent(in) :: x
         real :: f_result
      end function f
   end interface

   real, intent(in) :: a, b
   integer, intent(in) :: n
   real :: integral_result
   real :: h, total
   integer :: i

   h = (b - a) / n
!  Calculate the sum f(a)/2+f(a+h)+...+f(b-h)+f(b)/2
!  Do the first and last terms first
   total = 0.5 * (f(a) + f(b))
   do i = 1, n - 1
      total = total + f(a + i * h)
   end do

   integral_result = h * total
end function integral

end module integrate_module

program integrate

   use integrate_module

   implicit none
   intrinsic :: sin

   print *, integral(sin, a=0.0, b=3.14159, n=100)

end program integrate
```

Here is the result of running the program, which computes the integral of the trigonometric sine function from 0 to π. The intrinsic function `sin` is one that can be passed as an argument.

```
   1.9998353
```

3.15 Case Study: Calculating Probabilities I

Consider the problem of estimating the probability that a throw of two dice will yield a 7 or an 11. One way to solve this problem is to have a computer simulate many rolls of the dice and count how many times the result is 7 or 11. The probability of throwing 7 or 11 is then the number of successful throws divided by the total number of times the throw of the dice was simulated.

The Built-In Subroutine random_number

The heart of a probabilistic simulation program is a procedure that generates pseudo-random numbers. In Fortran, such a procedure is built in; it is a subroutine named `random_number`. The subroutine places uniformly distributed real numbers greater than or equal to 0 and less than 1 in the actual argument. The argument may be a single real variable or a real array. In this section, we will use `random_number` to generate one value at a time; in 4.7, we will use the same subroutine with an array as the argument to generate a whole array of random numbers with one subroutine call.

Computing the Probability of a 7 or 11

The program to estimate the probability of rolling 7 or 11 with two dice is built on a subroutine `random_int`, which in turn is based on the intrinsic subroutine `random_number`. To simulate the roll of one die, we need a subroutine that returns an integer from 1 to 6. The subroutine `random_int` has three arguments, `random_result`, `low`, and `high`. The first is used to store the result, which is, with approximately equal probability, any integer that is greater than or equal to `low`, the second argument, and that is less than or equal to `high`, the third argument. For example, the statement

```
call random_int(digit, 0, 9)
```

assigns to `digit` one of the 10 one-digit integers 0, 1, 2, ..., 9. `random_int` is written as a subroutine, rather than a function for two reasons:

1. It calls the subroutine `random_number`, which has the side effect of modifying the "seed" of the random generator; hence `random_int` itself has side effects. A function should not have a side effect.

2. If it were a function, it would be tempting to set the value of the variable `dice` with the statement

   ```
   dice = random_int(1, 6) + random_int(1, 6)
   ```

   An optimizing compiler might change this into the statement

   ```
   dice = 2 * random_int(1, 6)
   ```

   and each roll of the dice would produce an even number!

The program seven_11 simulates the event of rolling the dice 1000 times and computes a pretty good approximation to the true answer, which is 6/36 + 2/36 = 22.22%. On many systems (gfortran is an exception), calling the subroutine random_seed before generating any random numbers sets the generator to produce a different set of numbers each time the program is run. The seed also may be set using the put argument of random_seed.

```
module random_int_module

   implicit none
   private
   public :: random_int

contains

subroutine random_int(random_result, low, high)

   integer, intent(out) :: random_result
   integer, intent(in) :: low, high
   real :: uniform_random_value

   call random_number(uniform_random_value)
   random_result =  &
      int((high - low + 1) * uniform_random_value + low)

end subroutine random_int

end module random_int_module

program seven_11

   use random_int_module
   implicit none

   integer, parameter :: number_of_rolls = 1000
   integer :: die_1, die_2, dice, i, wins

   call random_seed()
   wins = 0
   do i = 1, number_of_rolls
      call random_int(die_1, 1, 6)
      call random_int(die_2, 1, 6)
      dice = die_1 + die_2
      if ((dice == 7) .or. (dice == 11)) then
         wins = wins + 1
      end if
   end do
```

```
   print "(a, f6.2)", &
      "The percentage of rolls that are 7 or 11 is", &
       100.0 * real(wins) / real(number_of_rolls)

end program seven_11
```

Here is the result of one execution of the program.

```
The percentage of rolls that are 7 or 11 is 22.40
```

Exercises

1. Write a program that determines by simulation the percentage of times the sum of two rolled dice will be 2, 3, or 12. You might want to use a `case` construct (2.3).

2. Two dice are rolled until a 4 or 7 comes up. Write a simulation program to determine the percentage of times a 4 will be rolled before a 7 is rolled. What was the largest sequence of rolls before the issue was decided?

3. Write a simulation program to determine the percentage of times exactly 5 coins will be heads and 5 will be tails, if 10 fair coins are tossed simultaneously.

4. Use the subroutine `random_int` to create a program that deals a five-card poker hand. Remember that the same card cannot occur twice in a hand. Use a character valued function `face_value(n)` that returns `"Ace"` for 1, `"2"` for 2, ..., `"10"` for 10, `"Jack"` for 11, `"Queen"` for 12, and `"King"` for 13, and another character-valued function `suit(m)` for the suit.

5. Modify the subroutine `random_int` so that the arguments `low` and `high` are optional. If `low` is not present, use the value 1. If `high` is not present, use the value `low` + 1. Test the subroutine with many different calls in which the optional arguments are omitted, arguments are called with keywords, and the arguments are in different orders.

3.16 Recursion

Recursion may be thought of as a mechanism to handle flow of control in a program, but its implementation requires dynamic storage allocation. Each time a recursive function or subroutine is called, there must be space for new copies of the variables that are local to the procedure. There is no way to tell at compile time how many times the routine will call itself; hence there is no way to determine the amount of storage needed to store copies of the variables local to a recursive procedure.

The use of recursion is a very powerful tool for constructing programs that otherwise can be quite complex, particularly if the process being modeled is described recursively. However, depending on the implementation available, recursion can require a substantial amount of runtime overhead. Thus, the use of recursion illustrates the classic trade-off between time spent in constructing and maintaining a program and execution time. In some cases, a process described recursively can be transformed into an iterative process in a very straightforward manner; in other cases, it is very hard and the resulting procedure is very difficult to follow. It is in these cases that recursion is such a valuable tool. We will illustrate some examples that fall into each category. A recursive version of the numerical integration program is discussed in 3.17. The sorting and selecting programs in 4.3 and 4.4 use recursion. It is used to evaluate expressions in 5.3 and to do exponentiation of big integers in 9.6. Recursion is used to process linked lists with pointers in 10.3 and to process trees with allocatable variables in 10.4.

The Factorial Function

First, let us look at the mathematical definition of the factorial function $n!$ defined for nonnegative integers. It is a simple example that will illustrate many of the important ideas relating to recursion.

$$0! = 1$$

$$n! = n \times (n-1)! \quad \text{for } n > 0$$

To use this definition to calculate 4!, apply the second line of the definition with $n = 4$ to get $4! = 4 \times 3!$. To finish the calculation we need the value of 3!, which can be determined by using the second line of the definition again. $3! = 3 \times 2!$, so that $4! = 4 \times 3 \times 2!$. Using the second line of the definition two more times yields $2! = 2 \times 1!$ and $1! = 1 \times 0!$. Finally, the first line of the definition can be applied to compute $0! = 1$. Plugging all these values back in produces the computation

$$4! = 4 \times 3 \times 2 \times 1 \times 1 = 24$$

From this, it is pretty obvious that an equivalent definition for $n!$ is

$$n! = n \times (n-1) \times (n-2) \times \cdots \times 3 \times 2 \times 1$$

for integers greater than zero. So this is an example for which it should be quite easy to write an iterative program as well as a recursive one; but to illustrate the recursive technique, let us first look at the recursive version. It should be easy to understand because it follows the recursive definition very closely.

```
module factorial_module

   implicit none
   public :: factorial

contains

recursive function factorial(n) result(factorial_result)

   integer, intent(in) :: n
```

```
   integer :: factorial_result

   if (n <= 0) then
      factorial_result = 1
   else
      factorial_result = n * factorial(n - 1)
   end if

end function factorial

end module factorial_module
```

The function is called using its name in an expression as shown by the simple program that computes 12!.

```
program test_factorial

   use factorial_module
   implicit none
   print*, "12! =", factorial(12)

end program test_factorial
```

```
 12! = 479001600
```

For a recursive function or subroutine, the keyword `recursive` must be placed on the procedure heading line. This version of the function returns a result of 1 for a negative value of `n` for which the mathematical factorial function $n!$ is undefined. Another alternative is to treat a negative argument as an error, but returning 1 keeps the example simple.

This program illustrates something often called **tail recursion**, which means that the only recursive call occurs as the very last step in the computation of the procedure. It is always easy to turn a process involving only tail recursion into an iterative process. Here is the iterative version of the factorial function.

```
function factorial(n) result(factorial_result)

   integer, intent(in) :: n
   integer :: factorial_result
   integer :: i

   factorial_result = 1
   do i = 2, n
      factorial_result = i * factorial_result
   end do

end function factorial
```

Note that the `do` loop will be executed zero times for any value of `n` that is less than 2, so that the value of 1 will be returned in these cases.

The Fibonacci Sequence

This next example illustrates not only the use of recursion when an iterative program would do as well, but a case in which a decision to implement a program based on a recursive definition yields an algorithm that has very poor running time, even if recursive function calls had no overhead.

The **Fibonacci sequence** 1, 1, 2, 3, 5, 8, 13, 21, 34, ... arises in such diverse applications as the number of petals in a daisy, the maximum steps it takes to recognize a sequence of characters, and the most pleasing proportions for a rectangle, the "golden section" of Renaissance artists and mathematicians. It is defined by the relations

$$f(1) = 1$$
$$f(2) = 1$$
$$f(n) = f(n-1) + f(n-2) \quad \text{for } n > 2$$

Starting with the third term, each Fibonacci number is the sum of the two previous Fibonacci numbers. Naive incorporation of the recurrence relation in a recursive function program is very easy, but produces an execution time disaster for all but the smallest values of n.

```
recursive function fibonacci(n) result(fibonacci_result)

   integer, intent(in) :: n
   integer :: fibonacci_result

   if (n <= 2) then
      fibonacci_result = 1
   else
      fibonacci_result = fibonacci(n - 1) + fibonacci(n - 2)
   end if

end function fibonacci
```

If the function is used to calculate $f(7)$, for example, the recursive calls request computation of $f(6)$ and $f(5)$. Then the computation of $f(6)$ again calls for the computation of $f(5)$ as well as $f(4)$. Thus, values of f are computed over and over with the same argument. In fact, the number of recursive function calls resulting from a single call to `fibonacci(n)` exceeds the answer, which is approximately $0.447 \times 1.618n$. The execution time of this function is called **exponential** because it depends on a number greater than 1 raised to the nth power.

To make this computation much more efficient, values of f must be saved and reused when needed, rather than being recomputed. The next function to compute the Fibonacci sequence is iterative rather than recursive. It uses the variables `f_i` and `f_i_minus_1` to hold the two most recently computed values of f and is iterative rather than recursive.

```
function fibonacci(n) result(fibonacci_result)

   integer, intent(in) :: n
   integer :: fibonacci_result
   integer :: f_i, save_f_i, i, f_i_minus_1

   if (n <= 2) then
      fibonacci_result = 1
   else
      f_i_minus_1 = 1
      f_i = 1
      do i = 3, n
         save_f_i = f_i
         f_i = f_i + f_i_minus_1
         f_i_minus_1 = save_f_i
      end do
      fibonacci_result = f_i
   end if

end function fibonacci
```

Although it may not be obvious at first glance why one must save the value of `f_i` in a variable `save_f_i` and only later copy it to `f_i_minus_1`, this function is by far more time and space efficient than the previous version. The speed increase is so marked that it is worth having a couple of lines of code that are not completely obvious.

The Towers of Hanoi

According to legend, there is a temple in Hanoi that contains a ritual apparatus consisting of 3 posts and 64 gold disks of graduated size that fit on the posts. When the temple was built, all 64 gold disks were placed on the first post with the largest on the bottom and the smallest on the top, as shown schematically in Figure 3-1. It is the sole occupation of the priests of the temple to move all the gold disks systematically until all 64 gold disks are on the third post, at which time the world will come to an end.

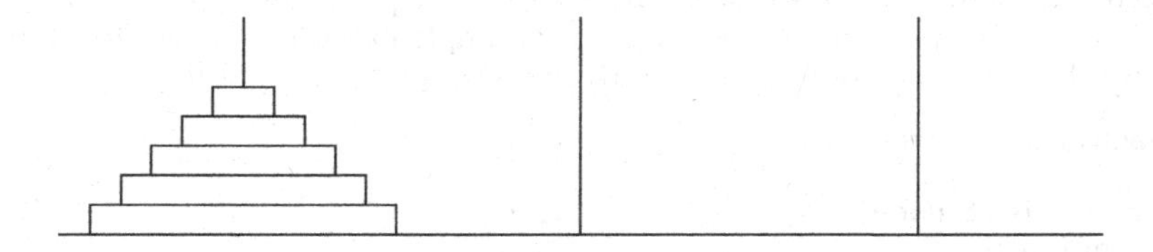

Figure 3-1 The towers of Hanoi

There are only two rules that must be followed:

1. Disks must be moved from post to post one at a time.

2. A larger disk may never rest on top of a smaller disk on the same post.

A smaller version of this apparatus with only eight disks made of plastic is sold as a recreational puzzle. The sequence of moves necessary to solve the simpler puzzle is not obvious and often takes hours to figure out. We propose to write a simple recursive procedure `hanoi` that prints complete directions for moving any number of disks from one post to another.

The recursive procedure `hanoi` is based on the following top-down analysis of the problem. Suppose n disks are to be moved from a starting post to a final post. Because the largest of these n disks can never rest on a smaller disk, at the time the largest disk is moved, all $n - 1$ smaller disks must be stacked on the free middle post as shown in Figure 3-2.

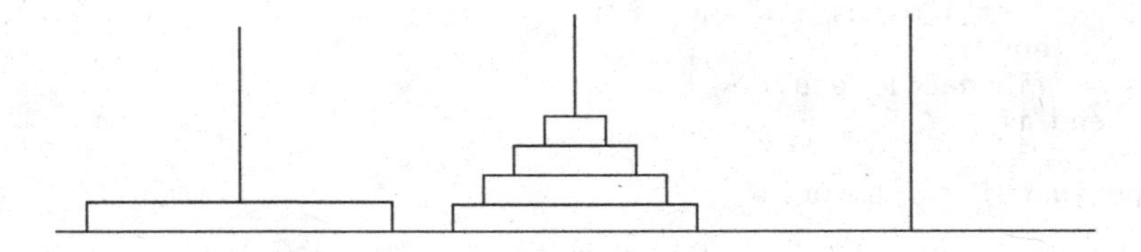

Figure 3-2 Locations of the disks when the largest disk is to be moved

For the number of disks $n > 1$, the algorithm has 3 steps.

1. Legally move the top $n - 1$ disks from the starting post to the free post.
2. Move the largest disk from the starting post to the final post.
3. Legally move the $n - 1$ disks from the free post to the final post.

The middle step involves printing a single move instruction. The first and third steps represent simpler instances of the same problem—simpler in this case because fewer disks must be moved. Therefore, the first and third steps may be handled by recursive procedure calls. In case $n = 0$, there are no instructions to be printed, and this provides a nonrecursive path through the procedure for the simplest case. The Fortran subroutine `hanoi`, its test program `test_hanoi`, and a sample execution output for four disks are shown. It is not easy to write an iterative version of this program.

```
module hanoi_module

   implicit none
   private
   public :: hanoi

contains

recursive subroutine hanoi(number_of_disks,  &
      starting_post, goal_post)
```

```
   integer, intent(in) ::  &
   number_of_disks, starting_post, goal_post
   integer :: free_post
   ! all_posts is the sum of the post values 1+2+3
   ! so that the free post can be determined
   ! by subtracting the starting_post and the
   ! goal_post from this sum.
   integer, parameter :: all_posts = 6

   if (number_of_disks > 0) then
      free_post =  &
      all_posts - starting_post - goal_post
      call hanoi(number_of_disks - 1,  &
                 starting_post, free_post)
      print *, "Move disk", number_of_disks,  &
            "from post", starting_post,  &
            "to post", goal_post
      call hanoi(number_of_disks - 1,  &
                 free_post, goal_post)
   end if

end subroutine hanoi

end module hanoi_module

program test_hanoi

   use hanoi_module
   implicit none

   integer :: number_of_disks

   read *, number_of_disks
   print *, "Input data  number_of_disks:",  &
         number_of_disks
   print *
   call hanoi(number_of_disks, 1, 3)

end program test_hanoi
```

```
 Input data  number_of_disks: 4

 Move disk 1 from post 1 to post 2
 Move disk 2 from post 1 to post 3
 Move disk 1 from post 2 to post 3
 Move disk 3 from post 1 to post 2
 Move disk 1 from post 3 to post 1
 Move disk 2 from post 3 to post 2
```

```
Move disk 1 from post 1 to post 2
Move disk 4 from post 1 to post 3
Move disk 1 from post 2 to post 3
Move disk 2 from post 2 to post 1
Move disk 1 from post 3 to post 1
Move disk 3 from post 2 to post 3
Move disk 1 from post 1 to post 2
Move disk 2 from post 1 to post 3
Move disk 1 from post 2 to post 3
```

Indirect Recursion

It is possible for procedures a and b to be indirectly recursive in the sense that a calls b and b calls a. An example of this kind of recursion occurs in the function `term` in 5.3.

Exercises

1. Write a recursive function `bc(n, k)` to compute the binomial coefficient $\binom{n}{k}$, $0 \le k \le n$, using the relations

$$\binom{n}{0} = 1$$
$$\binom{n}{n} = 1$$
$$\binom{n}{k} = \binom{n-1}{k-1} + \binom{n-1}{k} \quad \text{for } 0 < k < n$$

2. Write an efficient program to compute the binomial coefficient $\binom{n}{k}$.

3. The following recurrence defines $f(n)$ for all nonnegative integer values of n.

$$f(0) = 0$$
$$f(1) = f(2) = 1$$
$$f(n) = 2f(n-1) + f(n-2) - 2f(n-3) \quad \text{for } n > 2$$

   Write a function `f` to compute $f(n)$, $n \ge 0$. Also have your program verify that for $0 \le n \le 30$, $f(n) = [(-1)^{n+1} + 2^n] / 3$.

4. For positive integers a and b, the greatest common divisor of a and b satisfies the following recurrence relationship:

$$gcd(a,b) = b \quad \text{if } a \bmod b = 0$$

$$gcd(a,b) = gcd(b, a \bmod b) \quad \text{if } a \bmod b \ne 0$$

   Write a recursive function `gcd(a,b)` using these recurrences. Test the program by finding $gcd(24,36)$, $gcd(16,13)$, $gcd(17,119)$, and $gcd(177,228)$.

5. If the Towers of Hanoi procedure must be used to move n disks, how many individual moves must be made? If it takes one second for each move, how long will it take to move 64 disks? Hint: There are approximately 3.14×10^7 seconds per year.

3.17 Case Study: Adaptive Numerical Integration

To illustrate a very effective use of recursion to solve a problem of central importance in numerical computing, let us return to the program `integrate` from 3.14 that computes an approximation to a definite integral

$$\int_a^b f(x)\,dx$$

that represents the area bounded by the lines $x = a$, $x = b$, $y = 0$, and the curve $y = f(x)$, by a sum of the areas of n trapezoids, each of width h. The program used a function `integral` that takes arguments that are a function, the lower and upper limits of integration, and an integer that indicates the number of intervals to be used to form the approximating sum. The example in 3.14 computes

```
integral(sin, a=0.0, b=3.14159, n=100)
```

passing the function `sin` to be integrated as the first argument to the function `integral`.

Now suppose we want to integrate a function such as

$$f(x) = e^{-x^2}$$

Instead of using the previous version of the function `integral`, a slightly more sophisticated recursive function is used because decreasing the width of each trapezoid may not be the most efficient way to improve the accuracy of a trapezoidal approximation. In regions where the curve $y = f(x)$ is relatively straight, trapezoids approximate the area closely, and further reductions in the width of the trapezoids produce little further reduction in the error, which is already small. In regions where the curve $y = f(x)$ bends sharply, on the other hand, the area under the curve is approximated less well by trapezoids, and it would pay to concentrate the extra work of computing the areas of thinner trapezoids in such regions.

Another advantage of the function `integral` in this section is that it takes as input argument the maximum permitted error in the answer, rather than the number of subdivisions, whose relationship to the error in the answer is hard to predict in general.

The recursive function `integral` written in this section uses an **adaptive trapezoidal** method of approximating the area under a curve, requesting extra calculations through a recursive call only in those regions where the approximation by trapezoids is not yet sufficiently accurate.

Mathematicians tell us that the error $E(h)$ in approximating the area of the almost rectangular region with top boundary $y = f(x)$ by the area of one trapezoid is approximately $-1/12f''(c)h^3$, where h is the width of the trapezoid, and c is some x value in the interval, whose exact location may not be known, but which matters little because for reasonable functions $f''(x)$ varies little over a small interval of width h. The dependence of $E(h)$ on h^3 shows why the error drops rapidly as h decreases, and the dependence of $E(h)$ on $f''(c)$ shows why the error is smaller when $f''(x)$ is smaller, at places such as near inflection points (where the tangent line crosses the curve) where $f''(x) = 0$. If the same region is approximated by the sum of the areas of two trapezoids, each of width $h/2$, the error in each of them is approximately $-1/12f''(c_1)(h/2)^3$, or $1/8E(h)$, if we assume $f''(x)$ changes little over such a small interval so that $f''(c) \sim f''(c_1)$. Since there are two trapezoids, the total error $E(h/2)$ is approximately $E(h)/4$. If $T(h)$ and $T(h/2)$ are the two trapezoidal approximations and I is the exact integral, we have approximately

$$
\begin{aligned}
T(h/2) &= (I - E(h/2)) - (I - E(h)) \\
&= -E(h/2) + E(h) \\
&= -E(h/2) + 4E(h/2) \\
&= 3E(h/2)
\end{aligned}
$$

This formula provides a way to check whether the trapezoidal approximations are better than a specified error tolerance. Since

$$|E(h/2)| = \frac{1}{3}|T(h/2) - T(h)|$$

approximately, the two-trapezoid approximation is sufficiently accurate if

$$\frac{1}{3}|T(h/2) - T(h)| < tolerance$$

If not, then the error tolerance is split in two, and the adaptive trapezoidal function `integral` is called again to approximate the area of each half of the region to within half of the original error tolerance. Thus, only regions where the approximation error is still large are further subdivided.

```
module integral_module

   implicit none
   private
   public :: integral

contains

recursive function integral(f, a, b, tolerance)  &
      result(integral_result)

   intrinsic :: abs
   interface
      function f(x) result(f_result)
         real, intent(in) :: x
         real :: f_result
      end function f
   end interface
   real, intent(in) :: a, b, tolerance
```

```
   real :: integral_result
   real :: h, mid
   real :: one_trapezoid_area, two_trapezoid_area
   real :: left_area, right_area

   h = b - a
   mid = (a + b) /2
   one_trapezoid_area = h * (f(a) + f(b)) / 2.0
   two_trapezoid_area = h/2 * (f(a) + f(mid)) / 2.0 + &
                        h/2 * (f(mid) + f(b)) / 2.0
   if (abs(one_trapezoid_area - two_trapezoid_area)  &
         < 3.0 * tolerance) then
      integral_result = two_trapezoid_area
   else
      left_area = integral(f, a, mid, tolerance / 2)
      right_area = integral(f, mid, b, tolerance / 2)
      integral_result = left_area + right_area
   end if

end function integral

end module integral_module
```

To test the function `integral`, we write a small test program and a function subprogram `f`. The test program will evaluate

$$\int_{-4}^{4} e^{-x^2} dx$$

to an accuracy of 0.01. The curve $y = e^{-x^2}$ is an unnormalized error distribution function, used extensively in probability and statistics. Its integral is $\sqrt{\pi}$ (approximately 1.772454). It is assumed that `function_module` contains a function `f` that evaluates $f(x) = e^{-x^2}$.

```
program integrate

   use function_module
   use integral_module
   use math_module, only : pi
   implicit none

   real :: x_min, x_max
   real :: answer

   x_min = -4.0
   x_max = 4.0
   answer = integral(f, x_min, x_max, 0.01)
   print "(a, f11.6)",  &
```

```
         "The integral is approximately ", answer
   print "(a, f11.6)", &
         "The exact answer is           ", sqrt(pi)

end program integrate
```

```
The integral is approximately     1.777074
The exact answer is               1.772454
```

Because the modules `integral_module` and `math_module` both might be useful in contexts other than with this simple test program, it makes sense to keep them separate. When parts of a program are kept in separate files, the process of compiling and running the program could be a little more complicated, although how this is done depends on the system being used. In any case, it is important to ensure that the current version of each piece of the program is the one that is used. Many systems have programs, such as *make*, that help with this task.

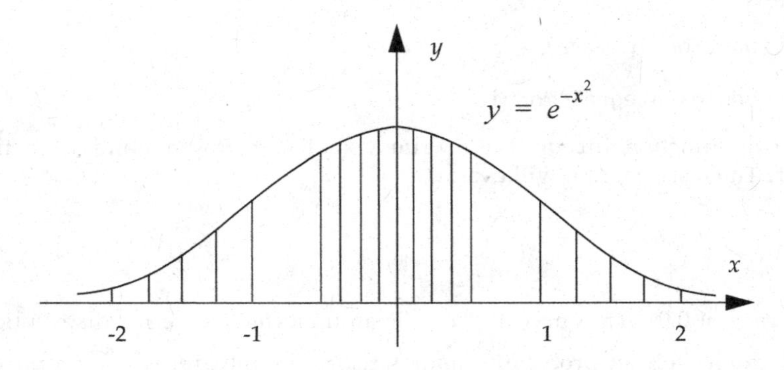

Figure 3-3 Approximating trapezoids used to calculate the integral of e^{-x^2}

Comparing the adaptive trapezoidal approximation to the exact answer, we see that the difference is approximately 0.0046, which is less than the specified error tolerance 0.01. Figure 3-3 shows the approximating trapezoids used between $x = -2$ and $x = +2$ to obtain the answer; trapezoids not shown have boundary points at $x = -4, -3, -2, 2, 3$, and 4. Notice that more trapezoids are required to keep within the error tolerance in the highly curved regions near the maximum of the function and where it first approaches zero than are required in the relatively straight regions near the two inflection points where the curve switches from concave upward to concave downward.

Exercises

1. Determine the number of trapezoids needed to evaluate

$$\int_{-4}^{4} e^{-x^2} dx$$

to an accuracy of 0.01 using the nonadaptive integration function discussed in 3.14.

2. Determine the approximate value of

$$\int_{0}^{2\pi} (e^x - \sin 2x) dx$$

using both the adaptive integration method of this section and the nonadaptive integration method discussed in 3.14. The area under the curve $y = f(x)$ between $x = a - h$ and $x = a + h$ may be approximated by the area under a parabola passing through the three points $(a - h, f(a - h))$, $(a, f(a))$, and $(a + h, f(a + h))$. The approximation, called Simpson's approximation, is given by the formula

$$\int_{a-h}^{a+h} f(x) dx = \frac{h}{3}[f(a-h) + 4f(a) + f(a+h)]$$

with error $-1/90f''''(c)h^5$ for some c in the interval of integration.

Use these facts to write a recursive adaptive Simpson's approximation function patterned on the adaptive trapezoidal approximation function `integral` in this section. Compare the number of recursive function calls for your adaptive Simpson's approximation function with the number required to achieve the same accuracy with the adaptive trapezoidal rule.

Arrays 4

In ordinary usage, a **list** is a sequence of values, usually all representing data of the same kind, or otherwise related to one another. A list of students registered for a particular course and a list of all students enrolled at a college are examples.

In Fortran, a collection of values of the same type is called an **array**. We will also refer to a one-dimensional array as a "list". Frequently, the same operation or sequence of operations is performed on every element in an array. On a computer that performs one statement at a time, it makes sense to write such programs by specifying what happens to a typical element of the array and enclosing these statements in a sufficient number of do constructs (loops) to make them apply to every element. Fortran also has powerful operations and intrinsic functions that operate on whole arrays or sections of an array. Programs written using these array operations are often clearer and are more easily optimized by Fortran compilers. Especially on computers with parallel or array processing capabilities, such programs are more likely to take advantage of the special hardware to increase execution speed.

4.1 Declaring and Using Arrays

We introduce the use of arrays with an example involving credit card numbers.

A Credit Card Checking Application

As an example of a problem that deals with a list, suppose that a company maintains a computerized list of credit cards that have been reported lost or stolen or that are greatly in arrears in payments. The company needs a program to determine quickly whether a given credit card, presented by a customer wishing to charge a purchase, is on this list of credit cards that can no longer be honored.

Suppose that a company has a list of 8262 credit cards reported lost or stolen, as illustrated in Table 4-1.

Since all of the 8262 numbers in the list must be retained simultaneously in the computer's main memory for efficient searching, and since a simple (scalar) variable can hold only one value at a time, each number must be assigned as the value of a variable *with a different name* so that the computer can be instructed to compare each account number of a lost or stolen card against the account number of the card offered in payment for goods and services.

Table 4-1 Lost credit cards

Account number of 1st lost credit card	2718281
Account number of 2nd lost credit card	7389056
Account number of 3rd lost credit card	1098612
Account number of 4th lost credit card	5459815
Account number of 5th lost credit card	1484131
.	.
.	.
.	.
Account number of 8262nd lost credit card	1383596

Subscripts

It is possible to use variables with the 8262 names

```
lost_card_1
lost_card_2
lost_card_3
  .
  .
  .
lost_card_8262
```

to hold the 8262 values. Unfortunately, the Fortran language does not recognize the intended relationship between these variable names, so the search program cannot be written simply. The Fortran solution is to declare a single object name `lost_card` that consists of many individual integer values. The entire collection of values may be referenced by its name `lost_card` and individual card numbers in the list may be referenced by the following names:

```
lost_card(1)
lost_card(2)
lost_card(3)
    .
    .
    .
lost_card(8262)
```

This seemingly minor modification of otherwise perfectly acceptable variable names opens up a new dimension of programming capabilities. All the programs in this chapter, and a large number of the programs in succeeding chapters, use arrays.

The numbers in parentheses that specify the location of an item within a list (or array) are **subscripts**, a term borrowed from mathematics. Although mathematical subscripts are usually written below the line (hence the name), such a form of typography is impossible on most computer input devices. A substitute notation, enclosing the subscript in parentheses or brackets, is adopted in most computer languages. It is customary to read the expression `x(3)` as "x sub 3", just as if it were written x_3.

The advantage of this method of naming the quantities over using the variable names `lost_card_1`, `lost_card_2`, ..., `lost_card_8262` springs from the following programming language capability: *The subscript of an array variable may itself be a variable,* or an even more complicated expression.

The consequences of this simple statement are much more profound than would appear at first sight.

For a start in describing the uses of a subscript that is itself a variable, the two statements

```
i = 1
print *, lost_card(i)
```

produce exactly the same output as the single statement

```
print *, lost_card(1)
```

namely, 2718281, the account number of the first lost credit card on the list. The entire list of account numbers of lost credit cards can be printed by the subroutine `print_lost_cards`.

```
subroutine print_lost_cards(lost_card)

   integer, dimension(:), intent(in) :: lost_card
   integer :: i

   do i = 1, 8262
      print *, lost_card(i)
   end do

end subroutine print_lost_cards
```

As an example of an array feature in Fortran, the collection of card numbers as a whole can be referenced by its name, so the `do` construct can be replaced by the one statement

```
print *, lost_card
```

The replacement just made actually creates a different output. The difference is that using the `do` loop to execute a `print` statement 8262 times causes each card number to be printed on a separate line. The new version indicates that as many as possible of the card numbers should be printed on one line, which might not produce acceptable output. Adding a simple format for the `print` statement instead of using the default produces a more desirable result, printing four card numbers per line.

```
print "(4i8)", lost_card
```

This is a little better, but another problem is that the number of lost and stolen cards varies daily. The subroutine will not be very useful if it makes the assumption that there are exactly 8262 cards to be printed each time. The declaration of an array-valued dummy argument indicates the number of subscripts, but does not fix the size of the dummy array.

```
integer, dimension(:), intent(in) :: lost_card
```

The colon indicates that the size of the array `lost_card` is to be assumed from the array that is the actual argument given when the subroutine is called. Also, this passed-on size can be used to print the entire list of cards using the intrinsic function `size`.

```
do i = 1, size(lost_card)
   print *, lost_card(i)
end do
```

The result would be a general subroutine for printing a list of integers.

Array Declarations

The name of an array must obey the same rules as an ordinary variable name. Each array must be declared in the declaration section of a program, module, or procedure. A name is declared to be an array by putting the `dimension` attribute in a type statement followed by a range of subscripts, enclosed in parentheses. For example,

```
real, dimension(1:9) :: x, y
logical, dimension(-99:99) :: yes_no
```

declares that `x` and `y` are lists of 9 real values and that `yes_no` is a list of 199 logical values. These declarations imply that a subscript for `x` or `y` must be an integer expression with a value from 1 to 9 and that a subscript for `yes_no` must be an integer expression whose value is from –99 to +99.

The lower bound may be omitted, in which case it is assumed to be 1. A declaration of x and y equivalent to the one above is

```
real, dimension(9) :: x, y
```

The **rank** of an array is the number of dimensions. The rank must not be greater than 15.

A parameter (named constant) array may have each upper bound be *, getting its shape from the constant expression that determines its value.

```
integer, parameter, dimension(0:*) :: perm = [ 5, 4, 3, 2, 1, 0 ]
```

In a function or subroutine, the range of a dummy argument usually consists of just the colon, possibly preceded by a lower bound, and the subscript range is determined by the corresponding actual argument passed to the procedure. This sort of dummy argument is called an **assumed-shape array**. If no lower bound is given, the subscript range is from 1 to the size of the array, in each dimension.

```
subroutine s(d)
   integer, dimension(:, :, 0:), intent(in) :: d
```

In this case, the subscripts on the dummy array d range from 1 to size(d,1) for the first subscript, from 1 to size(d,2) for the second, and from 0 to size(d,3)-1 for the third.

Style note: Declare dummy argument arrays to be assumed shape.

Arrays of character strings (5.1) may be declared like the following:

```
character(len=8), dimension(0:17) :: char_list
character(len=:). dimension(:,:), allocatable :: char_array
```

In this example, the variable char_list is a list of 18 character strings, each of length 8 and char_array is an allocatable rank-2 array.

The **shape** of an array is a list of the number of elements in each dimension. A 9×7 array has shape (9, 7); the array char_list declared above has shape (18); and the array declared by

```
integer, dimension(9, 0:99, -99:99) :: iii
```

has shape (9, 100, 199).

The shape of a scalar is a list with no elements in it. The shape of a scalar or array can be computed using the shape intrinsic function.

The declaration of a local array also may use values of other dummy arguments or values in its host (program or module) to establish extents and hence the shape of the array; such arrays are called **automatic arrays**. For example, the statements

```
subroutine s2(dummy_list, n, dummy_array)
   real, dimension(:) :: dummy_list
   real, dimension(size(dummy_list)) :: local_list
   real, dimension(n, n) :: dummy_array, local_array
   real, dimension(2*n+1) :: longer_local_list
```

declare that the size of dummy_list is to be the same as the size of the corresponding actual argument, that the array local_list is to be the same size as dummy_list, and that dummy_array and local_array are both to be two-dimensional arrays with n × n elements. The last declaration shows that some arithmetic on other dummy arguments is permitted in calculating array bounds; these expressions may include references to certain intrinsic functions, such as size and user-defined functions, in some circumstances.

If an array is declared outside a procedure, it must either be declared with constant fixed bounds or be declared to be allocatable or pointer and be given bounds by the execution of an allocate statement (see below) or a pointer assignment (10.1). In the first case, our lost and stolen card program might contain the declaration

```
integer, dimension(8262) :: lost_card
```

This is not satisfactory if the number of lost cards changes frequently. In this situation, one solution is to declare the array to have a sufficiently large upper bound so that there will always be enough space to hold the card numbers. Because the upper bound is fixed, there must be a variable whose value is the actual number of cards lost. Assuming that the list of lost credit cards is stored in a file connected to the standard input unit (unit=*), the following program fragment reads, counts, and prints the complete list of lost card numbers. The read statement has an iostat keyword argument whose value is set to zero if no error occurs and is set to a negative number if there is an attempt to read beyond the last data item in the file. In the program read_cards, the longer form of the read statement is required by the use of iostat.

```
program read_cards

   implicit none
   integer, dimension(20000) :: lost_card
   integer :: number_of_lost_cards, i, iostat_var

   do i = 1, 20000
      read (unit=*, fmt=*, iostat=iostat_var) lost_card(i)
      if (iostat_var < 0) then
         number_of_lost_cards = i - 1
         exit
      end if
   end do
      . . .
   print "(4i8)", lost_card(1:number_of_lost_cards)

end program read_cards
```

Although the array lost_card is declared to have room for 20,000 entries, the print statement limits output to only those lost card numbers that actually were read from the file by specifying a range of subscripts 1:number_of_lost_cards (see below for details about this notation).

Array Constructors

Rather than assign array values one by one, it is convenient to give an array a set of values using an array constructor. An **array constructor** is a sequence of scalar values defined along one dimension only. An array constructor is a list of values, separated by commas and delimited by the symbols "[" and "]". There are three ways to form the array constructor values and they may be combined in one constructor:

1. A scalar expression as in

```
x(1:4) = [ 1.2, 3.5, 1.1, 1.5 ]
```

2. An array expression as in

```
x(1:4) = [ a(i, 1:2), a(i+1, 2:3) ]
```

3. An implied do loop as in

```
x(1:4) = [ (sqrt(real(i)), i=1,4) ]
```

4. A combination of forms

```
x(1:5) = [ x(2), a(3, 1:2), (cos(x(i)), i=3,4) ]
```

If there are no values specified in an array constructor, the resulting array is zero sized. Unless there is a type specification (see below), the values of the components must have the same type and type parameters (kind and length). The rank of an array constructor is always one; however, the `reshape` intrinsic function can be used to define rank-2 to rank-15 arrays from the array constructor values. For example,

```
reshape ( [ 1, 2, 3, 4, 5, 6 ], [ 2, 3 ] )
```

is the 2 × 3 array $\begin{bmatrix} 1 & 3 & 5 \\ 2 & 4 & 6 \end{bmatrix}$

An **implied do list** is a list of expressions, followed by something that is like an iterative control in a `do` statement. The whole thing is contained in parentheses. It represents a list of values obtained by writing each member of the list once for each value of the `do` variable replaced by a value. For example, the implied `do` list in the array constructor above

```
(sqrt(real(i)), i=1,4)
```

is the same as the list

```
sqrt(real(1)), sqrt(real(2)), sqrt(real(3)), sqrt(real(4))
```

A `do` variable must be an integer variable and should be declared in the program or procedure where it is used. It must not be an array element. or a component of a structure (6.1). It must not have the `pointer` or `target` attribute (10.1). An implied `do` also can be used in an input/output list (11.3).

An array constructor may specify the type and type parameters of the resulting array. This allows a relaxation of the rule that says that all the values must have the same type and type parameters.

```
[ real :: 2, 3, 4.4 ] ! Type conversion

[ character(len=9) :: "Lisa", "Pamela", "Julie" ]

[ integer :: ]          ! An empty array
```

Dynamic Arrays

Dynamic storage allocation means that storage may be allocated or deallocated for variables during execution of the program. With dynamic storage allocation, the program can wait until it knows *during execution* exactly what size array is needed and then allocate only that much space. Memory also can be deallocated dynamically, so that the storage used for a large array early in the program can be reused for other large arrays later in the program after the values in the first array are no longer needed.

For example, instead of relying on an end-of-file condition when reading in the list of lost cards, it is possible to keep the numbers stored in a file with the number of lost cards as the first value in the file, such as

```
8262
2718281
7389056
1098612
5459815
1484131
   .
   .
   .
1383596
```

The program can then read the first number, allocate the correct amount of space for the array, and read the lost card numbers.

```
integer, dimension(:), allocatable :: lost_card
integer :: number_of_lost_cards
integer :: allocation_status
character(len=99) :: emsg
   . . .
! The first number in the file is the number
! of lost card numbers in the rest of the file.
read *, number_of_lost_cards
allocate (lost_card(number_of_lost_cards), &
      stat=allocation_status, errmsg=emsg)

if (allocation_status > 0) then
   print *, trim(emsg)
   stop
end if

! Read the numbers of the lost cards
read "(i7)", lost_card
   . . .
```

In the declaration of the array `lost_card`, the colon is used to indicate the rank (number of dimensions) of the array, but the bound is not pinned down until the `allocate` statement is executed. The `allocatable` attribute indicates that the array is to be allocated dynamically. Because the programmer does not know how many lost cards there will be, there is no way to tell the compiler that information. During execution, the system must be able to create an array of any reasonable size *after* reading from the input data file the value of the variable `number_of_lost_cards`.

If there is an allocation error (insufficient memory, for example), the variable `allocation_status` is set to a positive value, which may be tested by the programmer. If there is also an `errmsg` specifier, a message indicating the type of error is saved.

The component of a derived type (6.2) and a function result may be allocatable.

The **`deallocate` statement** may be used to free the allocated storage. Arrays declared allocatable in a procedure are deallocated when execution of the procedure is completed, if the array is allocated.

An allocatable array also may be allocated by an assignment statement (see the discussion under Array Assignment in this section).

Allocatable scalars will be used in later examples (10, 12.4).

Array Sections

In the following statement, used in the program `read_cards`, a section of the array `lost_card` is printed.

```
print "(4i8)", lost_card(1:number_of_lost_cards)
```

On many occasions, such as the one above, only a portion of the elements of an array is needed for a computation. It is possible to refer to a selected portion of an array, called an **array section**. A **parent array** is an aggregate of array elements, from which a section may be selected.

In the following example

```
real, dimension(10) :: a
   . . .
a(2:5) = 1.0
```

the parent array `a` has 10 elements. The array section consists of elements `a(2)`, `a(3)`, `a(4)`, and `a(5)`. The section is an array itself and the value 1.0 is assigned to all four of the elements in `a(2:5)`.

In addition to the ordinary subscript that can select a subobject of an array, there are two other mechanisms for selecting certain elements along a particular dimension of an array. One is a subscript triplet, and the other is a vector subscript.

The syntactic form of a **subscript triplet** is

[*expression*] : [*expression*] [: *expression*]

where each set of brackets encloses an optional item and each expression must produce a scalar integer value. The first expression gives a lower bound, the second an upper bound, and the third a stride. If the lower bound is omitted, the lower bound that was declared or allocated is used. (Note that an assumed-shape dummy array is treated as if it were declared with lower bound 1 unless a lower bound is given explicitly.) If the upper bound is omitted, the upper bound that was declared or allocated is used. If the declared bounds are `:`, the number of elements in each dimension is the size in that dimension. The **stride** is the increment between the elements in the section referenced by the triplet notation. If omitted, it is assumed to be one. For example, if `v` is a one-dimensional array (list) of numbers

```
v(0:4)
```

represents elements v(0), v(1), v(2), v(3), and v(4) and

```
v(3:7:2)
```

represents elements v(3), v(5), and v(7).

Each expression in the subscript triplet must be scalar. The values of any of the expressions in triplet notation may be negative. The stride must not be zero. If the stride is positive, the section is from the first subscript up to the second in steps of the stride. If the stride is negative, the section is from the first subscript down to the second, decrementing by the stride.

Another way of selecting a section of an array is to use a vector subscript. A **vector subscript** is an integer array expression of rank one. For example, if iv is a list of three integers, 3, 7, and 2, and x is a list of nine real numbers 1.1, 2.2, ..., 9.9, the value of x(iv) is the list of three numbers 3.3, 7.7, and 2.2—the third, seventh, and second elements of x.

Ordinary subscripts, triplets, and vector subscripts may be mixed in selecting an array section from a parent array. An array section may be empty.

Consider a more complicated example. If b were declared in a type statement as

```
real, dimension(10, 10, 5) :: b
```

then b(1:4:3, 6:8:2, 3) is a section of b, consisting of four elements:

```
b(1, 6, 3)   b(1, 8, 3)
b(4, 6, 3)   b(4, 8, 3)
```

The stride along the first dimension is 3; therefore, the notation references the first subscripts 1 and 4. The stride in the second dimension is 2, so the second subscript varies by 2 and takes on values 6 and 8. In the third dimension of b, there is no triplet notation, so the third subscript is 3 for all elements of the section. The section would be one that has shape (2, 2)—that is, it is two-dimensional, with extents 2 and 2.

To give an example using both triplet notation and a vector subscript, suppose again that b is declared as above:

```
real, dimension(10, 10, 5) :: b
```

then b(8:9, 5, [4, 5, 4]) is a 2 × 3 array consisting of the six values

```
b(8, 5, 4)   b(8, 5, 5)   b(8, 5, 4)
b(9, 5, 4)   b(9, 5, 5)   b(9, 5, 4)
```

If vs is a list of three integers, and vs = [4 5 4], the expression b(8:9, 5, vs) would have the same value as b(8:9, 5, [4, 5, 4]). The expression b(8:9, 5, vs) cannot occur on the left side of an assignment because of the duplication of elements of b.

The pack and unpack intrinsic functions may be useful in similar situations. As one simple example, the following program prints the positive elements of the array: 3, 7, and 4.

```
program print_pack

   implicit none
   integer, dimension(6) :: x = [3, -7, 0, 7, -2, 4]
   print *, pack(x, mask = (x > 0))

end program print_pack
```

Array Assignment

Array assignment assigns values to a collection of array elements. A simple example is

```
real, dimension(100, 100) :: a
   . . .
a = 0
```

Array assignment is permitted under three circumstances: when the array expression on the right has exactly the same shape as the array on the left, when the expression on the right is a scalar, and when the variable on the left is an allocatable array. In the first two cases, the expression on the right of the equals is **conformable** to the variable on the left. Note that, for example, if a is a 9 × 9 array, the section a(2:4, 5:8) is the same shape as a(3:5, 1:4), so the assignment

```
a(2:4, 5:8) = a(3:5, 1:4)
```

is valid, but the assignment

```
a(1:4, 1:3) = a(1:3, 1:4)
```

is not valid because. even though there are 12 elements in the array on each side of the assignment, the left side has shape (4, 3) and the right side has shape (3, 4).

When a scalar is assigned to an array, the value of the scalar is assigned to every element of the array. Thus, for example, the statement

```
m(k+1:n, k) = 0
```

sets the elements m(k+1, k), m(k+2, k), ..., m(n, k) to zero.

If the *name* of an allocatable array appears on the left side of an assignment statement, it is allocated to have the shape of the expression on the right and then assigned its value.

```
real, dimension(:), allocatable :: x1
   . . .
x1 = [1, 2, 3]
print *, size(x1)
x1 = [4, 5]
print *, size(x1)
```

prints the numbers 3 and 2.

Reading a List of Unknown Size

Dynamic array assignment allows a list, such as the list of lost credit cards, to be read in without knowing how many cards are in the list. Suppose the input file does not contain the number of cards, but is simply the list of lost credit cards.

```
2718281
7389056
1098612
5459815
1484131
   .
   .
   .
1383596
```

The following code reads in the numbers one at a time and allocates the array to be one bigger each time the new value is assigned as the last number in the array.

```
program read_cards_2

implicit none
integer, dimension(:), allocatable :: lost_card
integer :: card, ios
character(len=99) :: iom

lost_card = [ integer :: ]
do
   read (unit=*, fmt=*, iostat=ios, iomsg=iom) card
   if (ios < 0) exit
   if (ios > 0) then
      print *, trim(iom)
      cycle
   end if
   lost_card = [ lost_card, card ]
end do
!  . . .
print "(4i8)", lost_card

end program read_cards_2
```

Warning: if the list is very large, this program could run quite a long time, because the array `lost_card` is reallocated every time a new card is read as input. However, this can be fixed quite easily (see exercises below).

The where Construct

The **where construct** may be used to assign values to only those elements of an array where a logical condition is true; thus, it is often called a **masked array assignment**.

For example, the following statements set the elements of b and c to zero in those positions where the corresponding element of a is negative. The other elements of b and c are unchanged. a, b, and c must be arrays of the same shape.

```
where (a < 0)
   b = 0
   c = 0
end where
```

The logical condition in parentheses is an array of logical values conformable to each array in the assignment statement. In the example above, comparison of an array of values with a scalar produces the array of logical values.

The where construct permits any number of array assignments to be done under control of the same logical array. **elsewhere statements** within a where construct permit array assignments to be done where the logical expression is false and to indicate other conditions to affect additional statements. A where construct may contain nested where constructs.

The following statements assign to the array a the quotient of the corresponding elements of b and c in those cases where the element of c is not zero. In the positions where the element of c is zero, the corresponding element of a is set to zero and the zero elements of c are set to 1.

```
where (c /= 0) ! c/=0 is a logical array.
   a = b / c   ! a and b must conform to c.
elsewhere
   a = 0       ! The elements of a are set to 0
               ! where they have not been set to b/c.
   c = 1       ! The 0 elements of c are set to 1.
end where
```

The following program contains statements to set the array of integers key to −1, 0, or 1, depending on whether the corresponding element of the real array a is negative, zero, or positive, respectively. To see that the statements work correctly, the array a is filled with random numbers using the random_number subroutine. The values below the diagonal are negative; those above the diagonal are positive; and the diagonal is set to 0.

```
program elsewhere_example

implicit none
integer, parameter :: n=9
integer, dimension(n,n) :: key
integer :: i, j
real, dimension(n,n) :: a

call random_number(a)
do i=1, n
   do j = 1, n
      if (i > j) then
```

```
            ! Put negative numbers below the diagonal
            a(i,j) = -a(i,j) - 2.0
         else if (i < j) then
            ! Put positive numbers above the diagonal
            a(i,j) = a(i,j) + 2.0
         else
            ! Put zeros on the diagonal
            a(i,j) = 0.0
         end if
      end do
   end do

   where (a > 0)
      key = 1
   elsewhere (a < 0)
      key = -1
   elsewhere
      key = 0
   end where

   print "(9f5.1)", (a(i,:),i=1,9)
   print *
   print "(9i5)", (key(i,:),i=1,9)
   end program elsewhere_example
```

Here is the result of one execution of the program.

```
  0.0  2.6  2.8  2.3  2.1  2.5  2.3  2.5  2.3
 -3.0  0.0  2.6  2.9  2.4  2.4  2.8  2.6  2.9
 -2.3 -2.8  0.0  3.0  2.2  2.4  2.4  3.0  2.3
 -2.9 -2.3 -2.2  0.0  2.7  2.6  2.6  2.9  2.9
 -2.6 -2.2 -2.8 -2.3  0.0  2.1  2.5  2.5  2.5
 -2.1 -2.9 -2.5 -2.3 -2.1  0.0  2.7  2.8  2.3
 -2.3 -2.6 -2.9 -2.6 -2.9 -2.9  0.0  2.8  2.5
 -2.8 -2.8 -3.0 -2.7 -2.9 -2.2 -2.9  0.0  2.7
 -2.4 -2.1 -2.2 -2.1 -2.8 -2.8 -2.1 -2.6  0.0

    0    1    1    1    1    1    1    1    1
   -1    0    1    1    1    1    1    1    1
   -1   -1    0    1    1    1    1    1    1
   -1   -1   -1    0    1    1    1    1    1
   -1   -1   -1   -1    0    1    1    1    1
   -1   -1   -1   -1   -1    0    1    1    1
   -1   -1   -1   -1   -1   -1    0    1    1
   -1   -1   -1   -1   -1   -1   -1    0    1
   -1   -1   -1   -1   -1   -1   -1   -1    0
```

Within a `where` construct, only array assignments, nested `where` constructs, and `where` statements are permitted. The shape of all arrays in the assignment statements must conform to the shape of the logical expression following the keyword `where`. The assignments are executed in the order they are written—first those in the `where` block, then those in the `elsewhere` blocks.

Intrinsic Operators

All intrinsic operators and many intrinsic functions may be applied to arrays, operating independently on each element of the array. For example, the expression `abs(a(k:n))` results in a one-dimensional array of `n` – `k` + 1 nonnegative real values. A binary operation, such as *, may be applied only to two arrays of the same shape or an array and a scalar. It multiplies corresponding elements of the two arrays or multiplies the elements of the array by the scalar. The assignment statement

```
a(k, k:n+1) = a(k, k:n+1) / pivot
```

divides each element of `a(k, k:n+1)` by the real scalar value `pivot`. In essence, a scalar value may be considered an array of the appropriate size and shape with all its entries equal to the value of the scalar.

Element Renumbering in Expressions

An important point to remember about array expressions is that the elements in an expression may not have the same subscripts as the elements in the arrays that make up the expression. They are renumbered with 1 as the lower bound in each dimension. Thus, it is legal to add `y(0:7)` + `z(-7:0)`, which results in an array whose eight values are considered to have subscripts 1, 2, 3, ..., 8.

The renumbering must be taken into account when referring back to the original array. Suppose `v` is a one-dimensional integer array that is given an initial value with the declaration

```
integer, dimension(0:6), parameter :: v = [ 3, 7, 0, -2, 2, 6, -1 ]
```

The intrinsic function `maxloc` returns a list of integers giving the position (subscript) of the largest element of an array. `maxloc(v)` is [2] because position 2 of the list `v` contains the largest number, 7, even though it is `v(1)` that has the value 7. Also, `maxloc(v(2:6))` is the list [4] because the largest entry, 6, occurs in the fourth position in the section `v(2:6)`.

There is also an intrinsic function, `minloc`, whose value is the list of subscripts of a smallest element of an array. For example, if `a` = $\begin{bmatrix} 1 & 8 & 0 \\ 5 & -1 & 7 \\ 3 & -2 & 9 \end{bmatrix}$, the value of `minloc(a)` is [3 2] because `a(3, 2)` is the smallest element of the array.

Exercises

1. Write a statement that declares `values` to be an array of 100 real values with subscripts ranging from –100 to –1.

2. Use an array constructor to assign the squares of the first 100 positive integers to a list of integers named `squares`. For example, `squares(5)` = 25.

3. Write and run a program that declares an allocatable array named `squares`. Read an integer `n`. Assign to the array all of the perfect squares between 1 and n^2. Then write a loop to print out all the numbers from 1 to n^2 not in the array. Test the code with `n` = 8. One of the intrinsic functions `any` or `all` might be useful.

4. Declare a real rank-15 array `N` with default lower bounds and each upper bound 2. Assign the value 2 to each element of the array. Print the sum and product of all the values in the array.

5. Modify the program `read_cards_2` above to include an array `temp` of fixed size `chunk`=1000. Read card values into `temp` until it is full. Then assign the cards in `temp` as the last elements of the array `lost_card`. Empty the array `temp` and repeat the process until all the cards are read. Make sure to add the correct number of cards at the last step in case the number is not a multiple of `chunk`. Using the `cpu_time` intrinsic subroutine, compare the times for this program with the original `read_cards_2` program for reading 200,040 lost cards. Generate the file of lost cards with a program of some sort.

6. Suppose `list` is a one-dimensional array that contains `n` < `max_size` real numbers in ascending order. Write a subroutine `insert(list, n, max_size, new_entry)` that adds the number `new_entry` to the list in the appropriate place to keep the entire list sorted. Use a statement with array sections to shift the upper `k` elements one position higher in the array. See also the intrinsic functions `cshift` and `eoshift`. Do the same thing using a dynamic array assignment statement.

7. Write a function that finds the angle between two three-dimensional real vectors. If $v = (v_1, v_2, v_3)$, the magnitude of v is $|v| = \sqrt{v \cdot v}$, where $v \cdot v$ is the vector dot product of v with itself. The cosine of the angle between v and w is given by

$$\cos\theta = \frac{v \cdot w}{|v||w|}$$

The built-in function `acos` (arccosine) can be used to find an angle with a given cosine.

4.2 Searching a List

The previous section describes the appropriate terminology and some of the Fortran rules concerning arrays and subscripts. This section makes a start toward illustrating the power of arrays as they are used in meaningful programs. The application throughout this section is that of checking a given credit card account number against a list of account numbers of lost or stolen cards. Increasingly more efficient programs are presented here and compared.

The Problem: Credit Card Checking

When a customer presents a credit card in payment for goods or services, it is desirable to determine quickly whether it can be accepted or whether it previously has been reported lost or stolen or canceled for any other reason. The subroutines in this section perform this task. See 4.1 for ways to read the list `lost_card`.

Sequential Search through an Unordered List

The first and simplest strategy for checking a given credit card is simply to search from beginning to end through the list of canceled credit cards, card by card, either until the given account number is found in the list, or until the end of the list is reached without finding that account number. In the subroutine `search_1`, this strategy, called a **sequential search**, is accomplished by a `do` construct with exit that scans the list until the given account number is found in the list or all of the numbers have been examined.

Style note: It is good programming practice to make the searching part of the program a separate subroutine.

Other versions of the credit card program in this section will be obtained by modifying this subroutine.

The two ways of exiting from the search loop both pass control to the `end subroutine` statement. However, they have a different effect on the dummy argument `found`. When the credit card being checked is not in the list, the search loop is executed until the list is exhausted. This normal completion of the `do` construct allows control to fall through to the `end subroutine` statement with the value of `found` still false. When the card being checked is found in the list, the logical variable `found` is set to true before exiting the `do` construct. The calling program can test the actual argument passed to the dummy variable `found` to decide whether the card number was found in the list. The intrinsic function `size` used in this subroutine returns an integer value that is the number of elements in the array `lost_card`.

```
subroutine search_1(lost_card, card_number, found)

   integer, dimension(:), intent(in) :: lost_card
   integer, intent(in) :: card_number
   logical, intent(out) :: found
```

```
   integer :: i

   found = .false.
   do i = 1, size(lost_card)
      if (card_number == lost_card(i)) then
         found = .true.
         exit
      end if
   end do

end subroutine search_1
```

This subroutine makes a nice example for illustrating how individual elements of an array can be manipulated; but in Fortran, it is often better to think of operations for processing the array as a whole. In fact, using the built-in array functions, it is possible to do the search in one line.

```
found = any(lost_card == card_number)
```

The comparison

```
lost_card == card_number
```

creates a list of logical values with true in any position where the value of `card_number` matches a number in the list `lost_card`. The intrinsic function `any` is true if any of the elements in a list of logical values is true; it is false otherwise. The intrinsic function `any` may be thought of as an extension of the binary operator `.or.` to arrays.

The basic strategy of the program `search_1` is to check a credit card account number supplied as input against each account number, in turn, in the list of canceled or lost cards, either until a match is found or until the list is exhausted. These alternatives are not equally likely. Most credit cards offered in payment for purchases or services represent the authorized use of active, valid accounts. Thus, by far the most usual execution of the subroutine `search_1` is that the entire list is searched without finding the card number provided.

The number of comparisons a program must make before accepting a credit card is some measure of the efficiency of that program. For example, when searching for an acceptable credit card in a list of 10,000 canceled credit cards, the subroutine `search_1` usually makes 10,000 comparisons. On a traditional computer, the elapsed computer time for the search depends on the time it takes to make one comparison and to prepare to make the next comparison. However, on a computer with vector or parallel hardware, many comparisons may be done simultaneously and the intrinsic functions, probably written by the implementor to take advantage of this special hardware, might provide very efficient searching.

If the search must be performed on a traditional computer by making one comparison at a time, the search can be made more efficient by maintaining the list in the order of increasing card number. As soon as one canceled card number examined in the search is too large, all subsequent ones will also be too large, so the search can be abandoned early. The subroutine `search_2` presumes that the list is in increasing order.

```
subroutine search_2(lost_card, card_number, found)

   integer, dimension(:), intent(in) :: lost_card
   integer, intent(in) :: card_number
   logical, intent(out) :: found
   integer :: i

   found = .false.
   do i = 1, size(lost_card)
      if (card_number <= lost_card(i)) then
         found = (card_number == lost_card(i))
         exit
      end if
   end do

end subroutine search_2
```

Before accepting a presented account number, `search_1` always must search the entire list, but `search_2` stops as soon as it reaches a number in the list of canceled account numbers that is larger than or equal to the presented number.

Roughly speaking, the average number of comparisons needed for an acceptance by `search_2` is about half the list size, plus one additional comparison to determine whether the last entry examined was exactly the account number of the credit card being checked. For a list of 10,000 canceled cards, it would take an average of 5001 comparisons, significantly better than the 10,000 for `search_1`.

To a limited extent, this increased efficiency in the checking program is counterbalanced by some additional computer time needed to maintain the list of canceled credit cards in increasing order. However, the list is likely to be searched much more often than it is modified, so almost any increase in the efficiency of the checking program results, in practice, in an increase in the efficiency of the entire operation.

Program Notes

The sequential search loop in the subroutine `search_2` is not quite as straightforward as it seems at first glance. When the presented card `card_number` is compared against an entry `lost_card(i)`, three things can happen:

1. `card_number` is too high, in which case the search continues.

2. They match, in which case the presented card `card_number` has been found.

3. `card_number` is too low, in which case further search is futile.

The three possibilities are not equally likely. Case 1 can occur as many as 10,000 times in one search. Cases 2 and 3 can only happen once per search. It is important to test first for the most frequently occurring case. Otherwise, there will be two tests per iteration, slowing the search loop appreciably. This subroutine tests for the first case, and then, if it is false, determines whether case 2 or case 3 applies. The following `if` construct also does the tests in this same optimal order; but if the order of testing alternatives were changed, twice as many tests would be done.

```
found = .false.
do i = 1, size(lost_card)
   if (card_number > lost_card(i)) then
      cycle
   else if (card_number == lost_card(i)) then
      found = .true.
      exit
   else
      exit
   end if
end do
```

Binary Search

Sequential search is a brute force technique. It works well for short lists but is very inefficient for large ones. A somewhat different strategy, **divide and conquer**, is employed in a **binary search**. Half of the list can be eliminated in one comparison by testing the middle element. Then half the remaining elements are eliminated by another test. This continues until there is only one element left; then this element is examined to see if it is the one being sought. The list must be ordered for binary search. Note that a binary search is similar to what you do when looking up a telephone number in a phone book.

Table 4-2 shows how a binary search is used to try to find the number 2415495 in a list of 16 numbers. The numbers are given in increasing order in the first column. The presented number 2415495 is not in the list, but this fact plays no role in the search procedure until the very last step.

Table 4-2 A binary search that fails

Before any comparisons	After one comparison	After two comparisons	After three comparisons	After four comparisons	Given number
1096633	1096633				
1202604	1202604				
1484131	1484131				
1627547	1627547*				
2008553	2008553	2008553			
2202646	2202646	2202646*			
2718281	2718281	2718281	2718281*	2718281 ≠	2415495

Table 4-2 *(Continued)* A binary search that fails

Before any comparisons	After one comparison	After two comparisons	After three comparisons	After four comparisons	Given number
2980957*	2980957	2980957	2980957		
3269017					
4034287					
4424133					
5459815					
5987414					
7389056					
8103083					
8886110					

* An asterisk denotes the comparison entry at each stage, which is the last entry of the first half of the segment still under active consideration.

As a first step in binary searching, the list is divided in half. An asterisk follows the eighth number in column 1 because it is the last entry in the first half of the list. Since the given number 2415495 is less than (or equal to) the eighth entry 2980957, the second half of the list can be eliminated from further consideration. Column 2 shows only the first half of the original list (entries 1 through 8) retained as the segment still actively being searched.

The procedure is repeated. An asterisk follows the fourth entry in column 2 because it is the last entry in the first half of the segment of the list still actively being searched. Since the given number 2415495 is greater than the fourth number 1627547, this time it is the first half of the active segment that is eliminated and the second half (entries 5 through 8 of the original list) that is retained. This is shown in column 3 of Table 4-2.

In the next stage, the second remaining number 2202646, which was the sixth entry in the original list, is marked with an asterisk because it is the last entry of the first half of the segment still being searched. Since this number is exceeded by the given number 2415945, the second half of the segment in column 3 (entries 7 and 8) is retained as the active segment in column 4. The seventh entry of the original list, the number 2718281, is the last entry of the first half of the remaining list of two entries and thus is marked with an asterisk in column 4 to indicate its role as a comparison entry. Since the given number 2415495 is less than this, the other entry (the eighth original entry) is discarded, and column 5 shows that after four comparisons, only the seventh entry 2718281 remains as a candidate.

Since only one entry remains, a test for equality is made between the given number 2415495 and the one remaining entry 2718281. They are not equal. Thus, the given number is not in the list. Note that the previous comparisons of these two numbers were merely to determine whether the given number was less than or equal to the seventh entry.

Table 4-3 shows how the binary search works for the number 7389056, which is found in the list of 16 numbers. As before, the first column lists the original numbers with an asterisk following the last number of the first half of the list, the eighth entry. The number 7389056 is greater than the eighth entry, so the second half of the list (entries 9–16) is retained in column 2. A comparison of the given number 7389056 with the last entry of the first half of the segment remaining in column 2, the twelfth original entry 5459815, eliminates entries 9 through 12.

Table 4-3 A binary search that is successful

Before any comparisons	After one comparison	After two comparisons	After three comparisons	After four comparisons	Given number
1096633					
1202604					
1484131					
1627547					
2008553					
2202646					
2718281					
2980957*					
3269017	3269017				
4034287	4034287				
4424133	4424133				
5459815	5459815*				
5987414	5987414	5987414	5987414*		
7389056	7389056	7389056*	7389056	7389056 =	7389056
8103083	8103083	8103083			
8886110	8886110	8886110			

* An asterisk denotes the comparison entry at each stage, which is the last entry of the first half of the segment still under active consideration.

A comparison with the fourteenth entry, marked with an asterisk in column 3, eliminates the fifteenth and sixteenth entries. One more comparison of the given number 7389056 against the thirteenth entry, marked with an asterisk in column 4, eliminates that entry and leaves only the fourteenth entry 7389056. The final test for equality of the given number and the only remaining candidate in the list yields success, and it can be reported that the given number is the fourteenth entry in the list.

For the purpose of explanation, it is most convenient to use a list size that is an exact power of 2, that is, 2, 4, 8, 16, 32, ... This avoids fractions when the size of the list segment still under consideration is halved repeatedly. However, this is not essential; the use of integer division by 2 in the subroutine `binary_search` permits it to search a list of any length.

```
subroutine binary_search(lost_card, card_number, found)

   integer, dimension(:), intent(in) :: lost_card
   integer, intent(in) :: card_number
   logical, intent(out) :: found
   integer :: first, half, last, only

   first = 1
   last = size(lost_card)
   do
      if (first == last) exit
      half = (first + last) / 2
      if (card_number <= lost_card(half)) then
         ! Discard second half
         last = half
      else
         ! Discard first half
         first = half + 1
      end if
   end do

   ! The only remaining subscript to check is first
   ! (which is the same as last)
   only = first
   found = (card_number == lost_card(only))

end subroutine binary_search
```

When the part of the list still under consideration has been reduced to a single element by repeated bisection, the first element left is the last and only element left and the **do** construct is exited to test it.

Efficiency of a Binary Search

As before, we can get a reasonable indication of the efficiency of a search method by seeing how many times the given account number is compared against account numbers in the list of lost or stolen cards in the most usual event that the card number is not in the list.

The number of comparisons required in the binary search can be counted easily. With one data comparison, a list of items to be searched can be cut in half. When the list is reduced to one element, a final comparison determines whether that candidate is the credit card being searched for or not. Thus, with $n + 1$ comparisons, it is possible to search 2^n items. Turning it around the other way, n items may be searched using $\log_2 n+1$ comparisons. Thus, for example, 15 comparisons suffice for binary searching all lists of length up to 16,384 (= 2^{14}). This is considerably better than the 8192 comparisons needed for a sequential search! However, keep in mind that on a computer with intrinsic parallelism, it may be better to use the intrinsic functions and hope that the implementation takes advantage of the parallelism to do many comparisons simultaneously. Even if it does, whether or not it is faster than the binary search depends on the size of the list and the amount of parallelism in the system.

Exercises

1. What changes need to be made to the subroutine `binary_search` to search a list of integers with kind `long`?

2. How could the subroutine `search_2` be improved if you wanted to start at the end of the list when searching for a "large" number?

4.3 Sorting

Frequently it is necessary to sort a list of numbers or character strings. For example, the list `lost_card` in the previous section must be sorted for the binary search to work. One of the simplest ways to do this is to compare every number in the list with every other number in the list and swap them if they are out of order. As with the previous examples in this chapter, the sorting is done with a subroutine so that it can be put in a module and be used by many programs.

```
subroutine sort_1(list)

   real, dimension(:), intent(in out) :: list
   integer :: i, j

   do i = 1, size(list) - 1
      do j = i + 1, size(list)
         if (list(i) > list(j)) &
               call swap(list(i), list(j))
      end do
   end do

end subroutine sort_1
```

The subroutine `swap` of 3.5 that exchanges the values of two variables is assumed to be available, perhaps in a module. This is a very simple algorithm for sorting, but it is very inefficient and should not be used to sort more than a few hundred items.

A second approach to sorting a list is to find the smallest number in the list and put it in the first position, then find the smallest number in the remainder of the list and put it in the second position, etc. The built-in function `minloc` can be used effectively for this sort.

For an array `a` of rank n, that is, with n subscripts, the value of `minloc(a)` is a one-dimensional array whose entries are the n subscript positions of a smallest element of `a`. As described in 4.1, if the lower bound in a particular dimension is 1, the subscript position and the subscript value are the same. If not, the actual subscript can be found by adding the declared lower bound – 1 to the subscript position. In the subroutine `sort_2`, the subscript of `list(i:)` containing a minimal element is `min_loc(1)+i-1`, where `min_loc` is an array with one element used temporarily to store the value of `minloc(list(i:))`.

```
subroutine sort_2(list)

   real, dimension(:), intent(inout) :: list
   integer :: i
   integer, dimension(1) :: min_loc

   do i = 1, size(list) - 1
      min_loc = minloc(list(i:))
      call swap(list(i), list(i + min_loc(1) - 1))
   end do

end subroutine sort_2
```

This subroutine appears to be just about as inefficient as `sort_1`, because execution of the `minloc` function involves searching through the elements of `list(i:)` to find the smallest one. Indeed, it may be just as inefficient; however, if it is executed on a system with parallelism, the `minloc` function may be faster than a sequential search.

Quick Sort

One of the best sorting algorithms is called "quick sort" or "partition sort". Whereas `sort_1` needs to make approximately $n^2/2$ comparisons to sort n numbers, the quick sort needs approximately $n\log_2 n$ comparisons. To get an idea of the amount of improvement, for $n = 1000$ items, `sort_1` would require approximately 500,000 comparisons and the quick sort would require approximately 10,000 comparisons, a ratio of 50:1; for $n = 1{,}000{,}000$ items, `sort_1` would require approximately 500,000,000,000 comparisons and the quick sort would require approximately 100,000,000 comparisons, a ratio of 5000:1.

As might be expected, the quick sort is a bit more complicated. It is a divide-and-conquer algorithm like binary search. To sort a list of numbers, an arbitrary number (such as the first, last, or middle one) is chosen from the list. All the remaining numbers in turn are compared with the chosen number; the ones smaller are collected in a "smaller" set and the ones larger are collected in a "larger" set. The whole list is sorted by sorting the "smaller" set, following them with all numbers equal to the chosen number, and following them with the sorted list of "larger" numbers. Note that sorting the "smaller" and "larger" lists involves using the quick sort routine recursively (3.16).

```
recursive subroutine quick_sort(list)

   real, dimension(:), intent(in out) :: list
   real, dimension(:), allocatable :: smaller, larger
   integer :: i,  &
         number_smaller, number_equal, number_larger
   real :: chosen

   if (size(list) > 1) then
      allocate (smaller(size(list)))
      allocate (larger(size(list)))
      chosen = list(1)
      number_smaller = 0
      number_equal = 1
      number_larger = 0

      do i = 2, size(list)
         if (list(i) < chosen) then
            number_smaller = number_smaller + 1
            smaller(number_smaller) = list(i)
         else if (list(i) == chosen) then
            number_equal = number_equal + 1
         else
            number_larger = number_larger + 1
            larger(number_larger) = list(i)
         end if
      end do

      call quick_sort(smaller(1:number_smaller))
      list(1:number_smaller) =  &
            smaller(1:number_smaller)
      list(number_smaller+1:  &
            number_smaller+number_equal) = chosen
      call quick_sort(larger(1:number_larger))
      list(number_smaller+number_equal+1:) =  &
            larger(1:number_larger)
      deallocate (smaller, larger)
   end if

end subroutine quick_sort
```

Although the subroutine quick_sort follows the description fairly closely and sorts with order $n\log_2 n$ comparisons, it wastes a lot of space in each subroutine call creating new smaller and larger lists. However, by clever management of the available space (in fact, each element is replicated up to $\log_2 n$ times, creating a total memory use of $n\log_2 n$ reals), the entire list can be sorted without using any arrays except the original argument list itself. In the following version of the quick sort, the "smaller" numbers are collected together by placing them at the beginning of the list and the "larger" numbers are collected together by placing them at the end of the list. Also, every effort is made to eliminate unnecessary moving or swapping of elements in the list. To do serious sorting, this version should be used.

The details of the quick-sorting algorithm are still quite tricky and must be clarified further before an efficient and bug-free subroutine can be written. First, while it is possible to maintain two lists in a single one-dimensional array—the list `smaller` that grows up from the bottom of the array `list` and the list `larger` that grows down from the top of the array `list`—it is difficult to manage three lists in one array. Thus, the conditions for the sublists `smaller` and `larger` are relaxed to allow entries equal to the test element `chosen` to qualify for either of these sublists. Since these elements are the largest elements in the sublist `smaller`, and the smallest elements in the sublist `larger`, they are reunited in the middle of the array `list` when both sublists are sorted in place.

Second, since there are (essentially) no extra storage spaces for list elements, the only way to remove an unsuitably large element from the left (i.e., `smaller`) part of the list is to swap it with an unsuitably small element from the right (i.e., `larger`) part of the list. Each pass through the main loop of the subroutine `quick_sort` consists of a search for an unsuitably large element on the left, a search for an unsuitably small element on the right, and a swap.

If the input list is in completely random order, it does not matter which element of the list is chosen as the test element. We use the middle element of the input list for two reasons: (1) one of the more likely nonrandom orders of a list is that the list is already sorted; choosing the middle element as test element provides much better splits than the first or last in this case; (2) if the test element is the middle element, both the search in the left list for a "large" element and the search in the right list for a "small" element are guaranteed not to run off the ends of the list, because the middle element will stop both searches. A test for invalid subscripts can be eliminated from these two inner loops if the test element is the middle element.

The only argument to the subroutine `quick_sort` is the list of numbers to be sorted. Recall that within the subroutine, regardless of the lower and upper bound of the *actual* argument, the *dummy* argument has lower bound `1` and upper bound `n = size(list)`.

```
recursive subroutine quick_sort(list)

   real, dimension(:), intent(in out) :: list

   integer :: i, j, n
   real :: chosen, temp
   integer, parameter :: max_simple_sort_size = 6

   n = size(list)
   if (n <= max_simple_sort_size) then
      ! Use interchange sort for small lists
      call interchange_sort(list)
   else
      ! Use partition ("quick") sort
      chosen = list(n/2)
```

```
      i = 0
      j = n + 1

      do
         ! Scan list from left end
         ! until element >= chosen is found
         do
            i = i + 1
            if (list(i) >= chosen) exit
         end do
         ! Scan list from right end
         ! until element <= chosen is found
         do
            j = j - 1
            if (list(j) <= chosen) exit
         end do
         if (i < j) then
            ! Swap two out of place elements
            temp = list(i)
            list(i) = list(j)
            list(j) = temp
         else if (i == j) then
            i = i + 1
            exit
         else
            exit
         end if
      end do

      if (1 < j) call quick_sort(list(:j))
      if (i < n) call quick_sort(list(i:))
   end if ! test for small array

end subroutine quick_sort

subroutine interchange_sort(list)

   real, dimension(:), intent(in out) :: list
   integer :: i, j
   real :: temp

   do i = 1, size(list) - 1
      do j = i + 1, size(list)
         if (list(i) >  list(j)) then
            temp = list(i)
            list(i) = list(j)
            list(j) = temp
         end if
      end do
   end do
```

```
end subroutine interchange_sort
```

Sorting Small Lists

The subroutine quick_sort has been made more efficient by the addition of the following statements that test if the quantity of numbers to be sorted is small and call an interchange sort if it is.

```
if (n <= max_simple_sort_size) then
   ! Use interchange sort for small lists
   call interchange_sort(list)
```

Why be concerned about this? Quick sort rarely is used to sort such small lists, and even if it is, it is only relative efficiency that suffers: The absolute time required to quick sort a small list is very small. The answer is that although the user might not call quick_sort often to sort a very small list, because it is a divide-and-conquer technique, the quick-sort algorithm subdivides the list again and again until finally it calls itself recursively many times to sort very small lists. Thus, small inefficiencies in the quick sorting of small lists contribute many times over to form large inefficiencies in the quick sorting of large lists.

The solution is simple: for lists below a certain minimum size, interchange_sort is used. The subroutine quick_sort sorts all lists of size up to max_simple_sort_size using the compact and simple sorting algorithm of the subroutine sort_1 for such lists. For larger lists, it uses the quick sort algorithm. Some experimenting with randomly generated large lists and different values of max_simple_sort_size indicates that for this simple sorting algorithm and this implementation of the quick-sort algorithm, max_simple_sort_size = 6 is probably a good choice. Your mileage may vary (see Exercise 3).

Exercises

1. Modify the subroutine quick_sort so that a public variable named swap_count records the number of times two values are swapped. This provides a crude measure of the complexity of the sorting algorithm. Experiment with the program by generating 10,000 numbers using the built-in subroutine random_number discussed in 3.15. Also collect data about actual running time using the built-in subroutine cpu_time.

2. Execute quick_sort with randomly generated lists of numbers of various sizes of n between 16 and 32,768 to see if the number of values swapped is proportional to $n\log_2 n$.

3. Vary the parameter max_simple_sort_size and test quick_sort using randomly generated lists of size n = 10,000. What value of max_simple_sort_size produces the fewest swaps? Do not forget to count the swaps in interchange_sort. Does this value of max_simple_sort_size also produce the shortest actual running time? If time permits, see if the results change when n is increased to 100,000.

4.4 Selecting

A common problem is to find the median of a list of numbers, that is, the one that would be in the middle of the list if the list were in order. One way to do this is to sort the list and look at the element in the middle, but this is quite inefficient. The best sorting algorithms require $n\log_2 n$ steps to sort n numbers, whereas the median of n numbers can be found in n steps.

The trick is one that is often applicable to recursive procedures: Solve a slightly more general problem instead. In this case the more general problem to solve is to find the number that would be in position k, $1 \le k \le n$, if a list of n numbers were in order. Then to find the median, simply find the number in position $k = n/2$.

A good algorithm to select the kth element is similar to the quick-sort algorithm. Arbitrarily pick one of the numbers in the list. As with the quick sort, separate the numbers into three collections: the numbers smaller than the chosen number, the numbers equal to the chosen number, and the numbers larger than the chosen number. Suppose the size of each of these collections is s, e, and l, respectively. If $k \le s$, the number we are looking for is in the collection of smaller numbers, and, in fact, is the kth number in that collection in order; this number can be found by applying the same selection algorithm recursively to the list of smaller numbers. If $s < k \le s+e$, then the number chosen is the one we are looking for and the search is complete. If $s + e < k$, the number we are looking for is in the collection of larger numbers; it is, in fact, the one in position $k - s - e$ in that list in order, so it can be found by recursively calling the selection procedure.

Here is the Fortran program; the selected element is returned as the value of the variable element and the logical variable error indicates if a position outside the bounds of the list is requested. The procedure quick_select is written as a subroutine instead of a function because it returns two values.

```
recursive subroutine quick_select (list, k, element, error)

   real, dimension(:), intent(in) :: list
   integer, intent(in) :: k
   real, intent(out) :: element
   logical, intent(out) :: error
   real, dimension(:), allocatable :: smaller, larger
   integer :: i, n,  &
         number_smaller, number_equal, number_larger
   real :: chosen
```

```
   n = size(list)
   if (n <= 1) then
      error = .not. (n == 1 .and. k == 1)
      if (error) then
         element = 0.0  ! A value must be assigned
                 ! because element is intent(out)
      else
         element = list(1)
      end if

   else
      allocate (smaller(n), larger(n))
      chosen = list(1)
      number_smaller = 0
      number_equal = 1
      number_larger = 0

      do i = 2, n
         if (list(i) < chosen) then
            number_smaller = number_smaller + 1
            smaller(number_smaller) = list(i)
         else if (list(i) == chosen) then
            number_equal = number_equal + 1
         else
            number_larger = number_larger + 1
            larger(number_larger) = list(i)
         end if
      end do

      if (k <= number_smaller) then
         call quick_select  &
               (smaller(1:number_smaller),  &
                k, element, error)
      else if (k <= number_smaller + number_equal) then
         element = chosen
         error = .false.
      else
         call quick_select  &
               (larger(1:number_larger),  &
                k - number_smaller - number_equal,  &
                element, error)
      end if
      deallocate (smaller, larger)
   end if

end subroutine quick_select
```

Exercises

1. Modify the subroutine `quick_select` so that a variable named `compare_count` records the number of times two values are compared. This provides a crude measure of the complexity of the selection algorithm. Experiment with the program by generating 10,000 numbers using the built-in subroutine `random_number` (3.15). Also collect data about actual running time using the built-in subroutine `cpu_time` (8.2).

2. Execute `quick_select` with randomly generated lists of numbers of various sizes n to see if the number of values compared is proportional to n.

3. Rewrite `quick_select` to reduce the amount of temporary storage used, using the second version of `quick_sort` as a model.

4. Instead of using `list(1)` as the value of `chosen` in the subroutine `quick_select`, use `list(k)`. Repeat the timing experiments to see if this makes any difference. Try the experiments using both versions with a list that is already sorted.

4.5 Case Study: Solving Linear Equations

The operations of searching, sorting, and selecting discussed in the previous sections involve, by their nature, mostly operations on a single element of a list, one at a time. In many situations, particularly in numerical computations, whole arrays or sections of arrays can be processed at once. To explore an example of this type, we look at the problem of solving n simultaneous equations of the form

$$
\begin{aligned}
&a_{11}x_1 + a_{12}x_2 + \cdots + a_{1n}x_n = b_1 \\
&a_{21}x_2 + a_{22}x_2 + \cdots + a_{2n}x_n = b_2 \\
&\ldots \\
&a_{n1}x_n + a_{n2}x_2 + \cdots + a_{nn}x_n = b_n
\end{aligned}
$$

In matrix notation, this system of equations would be written as

$$
\begin{bmatrix} a_{11} & a_{12} & \cdots & a_{1n} \\ a_{21} & a_{22} & \cdots & a_{2n} \\ & \cdots & & \\ a_{n1} & a_{n2} & \cdots & a_{nn} \end{bmatrix}
\begin{bmatrix} x_1 \\ x_2 \\ \cdots \\ x_n \end{bmatrix}
=
\begin{bmatrix} b_1 \\ b_2 \\ \cdots \\ b_n \end{bmatrix}
$$

Solving the equations is done by performing combinations of the following operations, none of which changes the values of the solutions. The three operations are (1) interchanging equations (which amounts to interchanging rows in the matrix of coefficients), (2) multiplying an equation (i.e., row) by a constant, and (3) adding one equation (i.e., row) to another equation. The operations of interchanging columns in the matrix of coefficients (which amounts to renaming variables) and multiplying a column by a constant (which amounts to re-scaling the values of the variable represented by that column) are sometimes used in solving simultaneous linear equations, but are not used in the solution presented below.

These equations will be solved by a process called **Gaussian elimination.** Combinations of these operations are performed until the equations are in a form where all coefficients below the diagonal of the coefficient matrix are zero and all coefficients on the main diagonal are one; this constitutes the first phase of Gaussian elimination. In broad outline, what happens in this phase is that the first equation is solved for the first variable x_1 (i.e., its coefficient is made 1), and then appropriate multiples of the first equation are subtracted from each of the remaining equations to eliminate the variable x_1 from equations 2 to n. Then the second equation is solved for x_2, and multiples of it are subtracted from the remaining equations to eliminate x_2 also from equations 3 to n. Eventually, all the variables x_1, x_2, ..., x_{n-1} are eliminated from the nth equation, which can now be solved for x_n. At the end of the first phase, the set of equations takes the form

$$
\begin{aligned}
x_1 + c_{12}x_2 + c_{13}x_3 + \ldots + c_{1n-1}x_{n-1} + c_{1n}x_n &= d_1 \\
x_2 + c_{23}x_3 + \ldots + c_{2n-1}x_{n-1} + c_{2n}x_n &= d_2 \\
x_3 + \ldots + c_{3n-1}x_{n-1} + c_{3n}x_n &= d_3 \\
\ldots &= \ldots \\
x_{n-1} + c_{n-1n}x_n &= d_{n-1} \\
x_n &= d_n
\end{aligned}
$$

The second phase of Gaussian elimination is called **back substitution.** The last equation is already solved for $x_n = d_n$. The answer for x_n is substituted into the next-to-last equation, which contains only variables x_{n-1} and x_n after the first phase, so it can be solved for x_{n-1}. Then the answers for both x_n and x_{n-1} are substituted into the previous equation to solve for x_{n-2}, and so forth until all the variables x_n, x_{n-1}, ..., x_{n-2} are substituted into the first equation to solve for x_1.

An equivalent form of the back-substitution phase, which is used sometimes, is to subtract appropriate multiples of the nth equation from all previous equations to eliminate x_n from equations 1 to $n - 1$. Then multiples of equation $n - 1$ are subtracted from equations 1 to $n - 2$ to eliminate x_{n-1} from these equations. The process continues upward through the equations until each equation has only one variable, or equivalently, until every entry in the matrix of coefficients above the diagonal is zero. The equations now have the form

$$
\begin{aligned}
x_1 &= e_1 \\
x_2 &= e_2 \\
\ldots &= \ldots \\
x_n &= e_n
\end{aligned}
$$

which is solved for all of its variables. In the program solve_linear_equations, we use the first method, substituting directly without changing the triangular matrix of coefficients to this completely diagonalized form.

If all goes well, the process of solving the system of linear equations is no more complicated than what we just described; however, a general solution must foresee and provide for all possibilities, even the possibility that the set of equations is inconsistent and has no solution.

The first potential problem is that when we try to solve the first equation for the first variable x_1, we might find that the first equation does not involve x_1 (i.e., $a_{11} = 0$). If some other equation involves x_1, that is, if some $a_{k1} \neq 0$, then we can swap the first and kth equations (to make $a_{11} \neq 0$ after the swap) so that we can solve the new first equation for x_1 and proceed. On the other hand, if no equation involves x_1, then the system of equations does not uniquely determine x_1 and we must report this as an undetermined system of equations.

A similar problem might occur when we try to solve the kth equation for x_k. If the coefficient a_{kk} is zero at this point in the computation, then we must seek a later equation, say the mth, for which $a_{mk} \neq 0$, and swap it with the kth equation before proceeding. If all remaining coefficients in the kth column are zero, then x_k is not uniquely determined.

Conventional wisdom, which we follow in this program, says that even if a_{kk} is nonzero, it is still better to swap the kth equation with that later equation for which the absolute value $|a_{mk}|$ is largest. Part of the reason is that roundoff error in calculations with the real coefficients often results in a coefficient that should be zero being calculated as a small nonzero value, but almost never results in it being calculated as a large nonzero value. Swapping a_{kk} with a_{mk}, the coefficient with the largest magnitude, greatly reduces the risk of dividing by a coefficient a_{kk} that should have been calculated as zero.

The program solve_linear_equations makes heavy use of array operations and intrinsics to achieve compactness (and to illustrate the use of the operations and intrinsics). If these operations are not second nature, some statements in solve_linear_equations may require some puzzling out, perhaps by writing equivalent do constructs.

```
subroutine solve_linear_equations(a, x, b, error)

   real, dimension(:, :), intent(in) :: a
   real, dimension(:), intent(out) :: x
   real, dimension(:), intent(in) :: b
   logical, intent(out) :: error
   real, dimension(:, :), allocatable :: m
```

```
   integer, dimension(1) :: max_loc
   real, dimension(:), allocatable :: temp_row
   integer :: n, k

   error = size(a, dim=1) /= size(b) .or.  &
           size(a, dim=2) /= size(b)
   if (error) then
      x = 0.0
      return
   end if

   n = size(b)
   allocate (m(n, n+1), temp_row(n+1))
   m(1:n, 1:n) = a
   m(1:n, n+1) = b

   ! Triangularization phase
   triang_loop: do k = 1, n

      max_loc = maxloc(abs(m(k:n, k)))
      temp_row(k:n+1) = m(k, k:n+1)
      m(k, k:n+1) = m(k-1+max_loc(1), k:n+1)
      m(k-1+max_loc(1), k:n+1) = temp_row(k:n+1)

      if (m(k, k) == 0) then
         error = .true.
         exit triang_loop
      else
         m(k, k:n+1) = m(k, k:n+1) / m(k, k)
         m(k+1:n, k+1:n+1) = m(k+1:n, k+1:n+1) -  &
            spread(m(k, k+1:n+1), 1, n-k) *  &
            spread(m(k+1:n, k), 2, n-k+1)
      end if

   end do triang_loop

   ! Back substitution phase

   if (error) then
      x = 0.0
   else
      do k = n, 1, -1
         x(k) = m(k, n+1) - sum(m(k, k+1:n) * x(k+1:n))
      end do
   end if

   deallocate (m, temp_row)

end subroutine solve_linear_equations
```

The array `m` is created in the subroutine `solve_linear_equations` because the constant terms are subject to the same operations as the coefficients of the variables during the calculations of Gaussian elimination. It consists of the array a of coefficients enlarged by one column into which is placed the list of constants `b`. This is accomplished using the statements

```
real, dimension(:, :), allocatable :: m
n = size(b)
allocate (m(n, n + 1))
m(1:n, 1:n) = a
m(1:n, n+1) = b
```

Several array intrinsic functions are used in the subroutine `solve_linear_equations`. The `size` function is used to find the number of equations and variables, which is the size of the list `b`. The function `spread` takes an array and increases its dimension (i.e., number of subscripts) by one by duplicating entries along a chosen dimension. Suppose that `m` is the 3×4 array

$$\begin{bmatrix} 11 & 12 & 13 & 14 \\ 21 & 22 & 23 & 24 \\ 31 & 32 & 33 & 34 \end{bmatrix}$$

then `m(1, 2:4)` is the one-dimensional array

$$\begin{bmatrix} 12 & 13 & 14 \end{bmatrix}$$

and `spread(m(1, 2:4), dim=1, count=2)` is the two-dimensional array

$$\begin{bmatrix} 12 & 13 & 14 \\ 12 & 13 & 14 \end{bmatrix}$$

which consists of two copies of m(1, `2:4)` spread downward—that is, entries that differ only in the first subscript are duplicates. Similarly, `spread(m(2:3, 1), dim=2, count=3)` is the array

$$\begin{bmatrix} 21 & 21 & 21 \\ 31 & 31 & 31 \end{bmatrix}$$

consisting of three copies of `m(2:3, 1)` spread to the right, with duplicate entries that differ only in the second subscript. Since these two arrays are the same size and shape, they may be multiplied; the value of `spread(m(1, 2:4), 1, 2) * spread(m(2:3, 1), 2, 3)` is the array

$$\begin{bmatrix} 12 \times 21 & 13 \times 21 & 14 \times 21 \\ 12 \times 31 & 13 \times 31 & 14 \times 31 \end{bmatrix}$$

Thus, the resulting value of `m` after executing the statement

```
m(2:3, 2:4) = m(2:3, 2:4) -          &
        spread(m(1, 2:4), 1, 2) *    &
        spread(m(2:3, 1), 2, 3)
```

is

$$\begin{bmatrix} 11 & 12 & 13 & 14 \\ 21 & 22 - 12 \times 21 & 23 - 13 \times 21 & 24 - 14 \times 21 \\ 31 & 32 - 12 \times 31 & 33 - 13 \times 31 & 34 - 14 \times 31 \end{bmatrix}$$

It is not necessary to set `m(1,2:3)` to 0 to complete this step in the triangularization because these elements are never looked at again.

The intrinsic function `sum` finds the sum of all the elements of an array. If `a` is a one-dimensional array, then the statement

```
s = sum(a)
```

gives the same result (subject to rounding errors) as the statements

```
s = 0
do i = lbound(a), ubound(a)
   s = s + a(i)
end do
```

For higher-dimensional arrays, a nested `do` loop is needed for each dimension of the array to achieve the effect of the built-in function `sum`. Besides the added simplicity and clarity of using the expression `sum(a)` in place of nested loops, it is much easier for compilers to recognize that `sum(a)` applies the same operation to the entire array and therefore might be a suitable expression for parallel execution if the hardware permits. The `do` loop versions explicitly ask for the calculations to be done in a specific order and thus may not benefit from optimization.

Many other functions that operate on arrays are described briefly in 8.8 and Appendix A.

4.6 Case Study: Heat Transfer I

Consider the problem of determining the temperature at each point of a square plate when a heat source is applied to two edges and the heat has had time to be distributed throughout the plate.

For our purposes, we will assume that the plate consists of a 10 × 10 array of points. A constant source of heat with value 1.0 is applied to the left edge (column 1) of the plate and heat values 1.0, 0.9, 0.8, ..., 0.2, 0.1 are applied to the points at the top of the plate. No heat is applied to the other two borders. We assume that the temperature in the plate assumes a steady state when the temperature at each internal point is the average of the temperatures of the four points neighboring the point—the points to the north, east, west, and south. Thus, the program does an iterative calculation: at each step the temperature at each internal point is replaced by the average of the four surrounding points. This can be done as an array operation:

```
temp = (n + e + s + w) / 4.0
```

The `associate` construct is used to give short names to some of the sections of the two-dimensional array named `plate`. This makes it easier to understand that the main computational step is averaging the points to the north, east, south, and west.

Note also the use of parameters `tolerance` and `plate_format` in the program. The size of the plate is also a parameter `P` so that it can be changed easily.

```
!  A simple solution to the heat transfer problem
!     using arrays and the associate construct

program heat

   implicit none
   integer, parameter :: P = 10
   real, dimension(P, P), target :: plate
   real, dimension(P-2, P-2)          :: temp
   real, parameter :: tolerance = 1.0e-4
   character(len=*), parameter :: plate_format = "(10f5.2)"

   real    :: diff
   integer :: i,j, niter

   ! Set up initial conditions
   plate = 0
   plate(:, 1) = 1.0   ! boundary values
   plate(1, :) = [ ( real(j)/P, j = P, 1, -1 ) ]

   !  Alias parts of the plate
   associate (inside => plate(2:P-1, 2:P-1), &
                   n => plate(1:P-2, 2:P-1), &
                   s => plate(3:P,   2:P-1), &
                   e => plate(2:P-1, 1:P-2), &
                   w => plate(2:P-1, 3:P))

      ! Iterate
      niter = 0
      do
         temp = (n + e + s + w) / 4.0
```

```
         diff = maxval(abs(temp-inside))
         niter = niter + 1
         inside = temp
         print *, niter, diff
         if (diff < tolerance) exit
      end do
   end associate

   do i = 1, min(P, 10)
     print plate_format, plate(i, :)
   enddo

end program heat
```

Here are the results produced by the last print statement after the computation has converged.

```
1.00 0.90 0.80 0.70 0.60 0.50 0.40 0.30 0.20 0.10
1.00 0.89 0.78 0.67 0.57 0.47 0.36 0.26 0.14 0.00
1.00 0.88 0.76 0.65 0.54 0.43 0.33 0.23 0.12 0.00
1.00 0.87 0.74 0.62 0.50 0.40 0.30 0.20 0.10 0.00
1.00 0.85 0.71 0.58 0.46 0.36 0.26 0.18 0.09 0.00
1.00 0.82 0.66 0.52 0.41 0.31 0.23 0.15 0.07 0.00
1.00 0.77 0.59 0.45 0.34 0.25 0.18 0.12 0.06 0.00
1.00 0.69 0.48 0.34 0.25 0.18 0.13 0.08 0.04 0.00
1.00 0.49 0.29 0.19 0.13 0.10 0.07 0.04 0.02 0.00
1.00 0.00 0.00 0.00 0.00 0.00 0.00 0.00 0.00 0.00
```

4.7 Case Study: Calculating Probabilities II

In 3.15, we considered the problem of calculating the probability that a throw of two dice will yield a 7 or an 11. The resulting program used the built-in subroutine random_number to generate a random number between 0 and 1. We now provide a slightly different solution in which the main program uses arrays.

Generating an Array of Random Numbers

In this section, we will rewrite the subroutine random_int to return an array of integers from low to high. The subroutine random_int calls random_number, but this time it passes an array to be filled with numbers from 0 to 1.

```
subroutine random_int(result, low, high)

   integer, dimension(:), intent(out) :: result
   integer, intent(in) :: low, high
   real, dimension(:), allocatable :: uniform_random_value
```

```
      allocate (uniform_random_value(size(result)))
      call random_number(uniform_random_value)
      result = int((high - low + 1) * uniform_random_value + low)
      deallocate (uniform_random_value)

   end subroutine random_int
```

Computing the Probability of a 7 or 11 Using Arrays

Using the array version of the subroutine random_int, the program to estimate the probability of rolling 7 or 11 with two dice is a bit shorter than the scalar version. We leave it to the reader to ponder whether it is easier or more difficult to understand than the scalar version.

```
program seven_11

   implicit none
   integer, parameter :: number_of_rolls = 1000
   integer, dimension (number_of_rolls) :: dice, die_1, die_2
   integer :: wins

   call random_seed()
   call random_int (die_1, 1, 6)
   call random_int (die_2, 1, 6)
   dice = die_1 + die_2
   wins = count ((dice == 7) .or. (dice == 11))

   print "(a, f6.2)",  &
      "The percentage of rolls that are 7 or 11 is", &
      100.0 * real (wins) / real (number_of_rolls)

contains

subroutine random_int . . .
   . . .

end program seven_11
```

The built-in function count returns the number of true values in any logical array; in this case the value in the array is true if the corresponding value in the array dice is 7 or 11. This version of the program seven_11 should produce an answer similar to the one produced by the scalar version.

Exercises

1. Use random_int to write a program that determines by simulation the percentage of times the sum of two rolled dice will be 2, 3, or 12.

2. Two dice are rolled until a 4 or 7 comes up. Use `random_int` to write a simulation program to determine the percentage of times a 4 will be rolled before a 7 is rolled.

3. Use `random_int` to write a simulation program to determine the percentage of times exactly 5 coins will be heads and 5 will be tails, if 10 fair coins are tossed simultaneously. If you use a two-dimensional array, it might help to make `random_int` an impure elemental subroutine (8.7).

4. Is it reasonable to use the array version of `random_int` with an array argument to create a program that deals a five-card poker hand? Remember that the same card cannot occur twice in a hand.

5 Character Data

In a computer program, a piece of written text is called a **character string**. Character strings have been used throughout this book to retain messages and identify information printed out but not processed in any significant way. This chapter reviews this simple use of character strings and presents computer programs in which the character strings themselves are the center of interest.

5.1 Use of Character Data in Fortran Programs

Character String Declarations

A character string variable in a Fortran program is declared to be type character. Each object of type character has a length, which is the number of characters that the string has. For example, the declaration

```
character(len=7) :: string_7
```

declares the variable `string_7` to be a character string of length 7. If the length is omitted, it is assumed to be 1.

Character dummy arguments and character parameters may have their length designated as an asterisk, indicating that their length will be determined by the corresponding actual argument or constant.

> *Style note:* Declare dummy argument and parameter character strings to be length (*).

A character string may be declared to have its length determined when it is assigned or allocated. In this case the length is : and the variable must be declared to be allocatable.

```
character(len=:), allocatable :: line_4
```

It is possible for a character string to have length zero. It is not particularly useful to declare a variable to have length zero because such a variable could only assume one value, called the **null string**. However, the null string can arise as a result of a computation and such a result might need to be saved as the value of a variable.

The characters in a character string are numbered 1 to n, where n is the length of the string.

It is possible to have an array of character strings, all of the same length. The following declares `string_array` to be a 5 × 9 × 7 array of character strings of length 20.

```
character(len=20), dimension(5, 9, 7) :: string_array
```

Character Parameters

A character constant may be given a name using the **parameter** attribute. The program `hello` prints a character parameter or named character constant, instead of a literal character constant.

```
program hello
   implicit none
   character(len=*), parameter :: &
         message = "Hello, I am a computer."
   print *, message
end program hello
```

Running the program produces

```
Hello, I am a computer.
```

Note that the name of the character parameter must be declared, just like a character variable, but the length may be an asterisk indicating that the length is to be determined from the value of the string.

Another excellent use of a character parameter is as a format in an input/output statement.

Character Constants

Recall that a character constant is enclosed in quotation marks (double quotes). This makes it possible for the computer to tell the difference between the character constant `"yes"` and the variable `yes`, or between the character constant `"14"` and the integer constant `14`.

Assigning Values to Character Variables

A variable that has been declared to be a character string may be assigned a value that is a character string. A simple example is provided by the following program that assigns a string to a character variable used in a `print` statement instead of executing alternative `print` statements containing different messages.

```
program test_sign
   implicit none
   real :: number
   character(len=:), allocatable :: number_sign

   read *, number
```

```
   if (number > 0) then
      number_sign = "positive"
   else if (number == 0) then
      number_sign = "zero"
   else
      number_sign = "negative"
   end if
   print *, number, "is ", number_sign
end program test_sign
```

A sample output is:

```
-2.3000000 is negative
```

Length of a Character String

The **length** of a character string is the number of characters in the string. The length of a character string is never negative. Each blank occurring in the string is counted in its length. The built-in function `len` gives the length of a character string. Thus,

`len("love")` is 4
`len("Good morning.")` is 13
`len(" ")` is 1
`len("    ")` is 4
`len("bg7*5 ad")` is 8

As with other functions, the argument of the function `len` may be a variable or more general expression, as well as a constant.

For a character string declared to have a fixed length, the length of a character string assigned to a character variable may be different from the length declared for that variable. For example, if the input number is zero in the program `test_sign`, the four-character constant "zero" is assigned to the eight-character variable `number_sign`. This assignment is legal. Four blanks are added to the end of the string `zero` to make its length 8, the declared length of the variable `number_sign`. Thus, the new value of the variable `number_sign` is "zero*bbbb*", where *b* is a blank character.

On the other hand, if a fixed-length character string to be assigned to a variable is longer than the declared length of the variable, characters are truncated from the right end of the string prior to assignment. For example, if the string `name` has a declared length of 3, the assignment statement

```
name = "Jonathan"
```

results in the string "Jon" being assigned to `name`.

Most of the difficulties created by these assignment rules are avoided by declaring a `number_sign` to have length :. Then when it is assigned the four-character value "zero", its length is four. If it is assigned the value "negative", its length is 8.

In a subprogram, the length of a character dummy argument may be given as an asterisk (*), which means that the length of the corresponding actual argument is to be used. Such a dummy argument is said to have **assumed length**. For example,

```
subroutine process(c)
   character(len=*), intent(in) :: c
```

The length of a local character string may depend on values related to the dummy arguments; such strings are called **automatic**, as they are very similar to automatic arrays (4.1).

For example, a temporary local string can be declared to hold the value of an argument passed in.

```
subroutine ss(c)
   character(len=*), intent(in) :: c
   character(len=len(c)) :: temp_c
   temp_c = c
      . . .
```

The intrinsic function `len` provides information that is otherwise unobtainable in the case where a character string, such as the variable `c` above, is a dummy argument with its length given by an asterisk or the case where the string is declared with length `:`.

Trimmed Length of a String

In some cases, we are not interested in the length in characters of a variable. A definition of length that is suitable for many applications is the length of the substring that includes all characters up to and including the last nonblank character, but excluding terminal blanks. The function `len_trim` is a built-in function that computes the length of a trimmed string. In addition, there is a built-in function `trim`, whose value is the given character string with all trailing blanks removed.

Input of Character Strings

When character strings are supplied as input data for a `read` statement with the default format (*), the string should be enclosed in quotes, just like a character constant. When using an a format, however, surrounding quotes must be omitted; any quotes among the characters read are considered to be part of the character constant.

Character Collating Sequences

Most Fortran programming language systems use the standard 128-character ASCII character set. The acronym ASCII stands for "American Standard Code for Information Interchange"; however, essentially the same code is also the international standard ISO/IEC 646:1991. Some systems also may support a newer standard ISO/IEC 10646:-1:2000, a multi-byte code designed to handle characters from many languages. The remainder of this section assumes the character encoding is ASCII.

The intrinsic ordering for characters, called the **collating sequence**, is shown in Table 5-1 for a selection of printable ASCII characters. One character is considered "less than" another character if it precedes the other character in the collating sequence.

Table 5-1 The collating sequence for printable ASCII characters

blank ! " # $ % & ' ( ) * + , - . /

0 1 2 3 4 5 6 7 8 9 : ; < = > ? @

A B C D E F G H I J K L M N O P Q R S T U V W X Y Z [\ ] ^ _ `

a b c d e f g h i j k l m n o p q r s t u v w x y z { | } ~

The Built-In Functions `iachar` and `achar`

The built-in function `iachar` produces an integer representing the ASCII code of the character given as argument. For example, `iachar("A")` is 65.

The function `achar` returns the character with a given code. For example, `achar(65)` gives the ASCII character in position 65, which is "A".

A Testing Technique for Character Output

The program `explore_character_set` will allow you to explore the ASCII collating sequence one character at a time. You type the character code and the computer prints the character with that code. It should be run interactively.

```
program explore_character_set

! Prints the character with given character code
! in the default kind

   implicit none
   integer :: code

   print *, "Type a character code"
   read *, code
   print "(i5, 3a)", code, ">", achar(code), "<"

end program explore_character_set
```

```
 Type a character code
65
   65>A<
```

The blank character is a perfectly valid character (ASCII code 32). To better see the value of `achar`, the value is printed surrounded by the printable characters > and <. A blank character will then conspicuously occupy the print or display position between its delimiters.

You must expect some surprises when you run the program explore_character_set. Most of the characters from 0 to 31 do not print. Some, like the line feed or new line, achar(10) in ASCII, direct the printer to perform some action rather than print a character. The delimiters > and < will help you figure out what action was taken.

Table 5-2 summarizes executions of the program explore_character_set. It is the output of a program similar to explore_character_set that uses loops to show the printable ASCII characters and their corresponding codes, eight per line of output. Codes 0 through 31 and code 127 represent special control characters such as the "bell" character, backspace, and newline. Code 32 represents the blank character.

Table 5-2 The printable ASCII characters

32	33 !	34 "	35 #	36 $	37 %	38 &	39 '
40 (	41)	42 *	43 +	44 ,	45 -	46 .	47 /
48 0	49 1	50 2	51 3	52 4	53 5	54 6	55 7
56 8	57 9	58 :	59 ;	60 <	61 =	62 >	63 ?
64 @	65 A	66 B	67 C	68 D	69 E	70 F	71 G
72 H	73 I	74 J	75 K	76 L	77 M	78 N	79 O
80 P	81 Q	82 R	83 S	84 T	85 U	86 V	87 W
88 X	89 Y	90 Z	91 [	92 \	93]	94 ^	95 _
96 `	97 a	98 b	99 c	100 d	101 e	102 f	103 g
104 h	105 i	106 j	107 k	108 l	109 m	110 n	111 o
112 p	113 q	114 r	115 s	116 t	117 u	118 v	119 w
120 x	121 y	122 z	123 {	124 |	125 }	126 ~	

Comparison of Character Strings

The comparison operators

```
<, <=, ==, /=, >, >=
```

may be used to compare character values.

The ordering of strings is an extension of the ordinary lexicographic (i.e., dictionary) ordering of words, but uses the processor codes to order characters other than letters. If the first character of one character string precedes the first character of the second string in the collating sequence, then we say the first character string is less than the second. If the first characters are equal, the second characters are used to decide which character string is smaller. If the second characters also match, the third characters are used to decide, and so on. The character string with the smaller character in the first position where the two strings differ is considered the smaller character string. When character strings of different lengths are compared, the shorter one is treated as if it were padded with enough blanks at the end to make it the same length as the longer one. For example,

```
"apple" < "bug" < "cacophony" < "doldrums"

"earache" < "elephant" < "empathy" < "equine"

"phlegmatic" < "phonograph" < "photosynthetic"

"dipole" < "duplicate" == "duplicate   " < "dynamic"
```

In the first line of expressions, decisions are made on the basis of the first letter of the strings. In the second line, since each string has first letter "e", decisions are made on the basis of the relative collating position of the second letters. In the third set of comparisons, third or fourth letters differ.

From these examples, it is clear that the natural order of character strings corresponds exactly to ordinary alphabetic order when the character strings are words written either entirely in lowercase or entirely in uppercase letters.

String ordering does not take meaning into account. For example, although

```
"1" < "2" < "3" < "4"
```

as expected, it is also true that

```
"four" < "one" < "three" < "two"
```

and, worse yet

```
"12" < "2"
```

String ordering is also sensitive to upper and lower case. The two character strings

```
"word"        "WORD"
```

are not equal.

Substrings

Many character-processing applications require breaking down a string into individual characters or subsequences of characters. Examples are decomposing a word into letters or a sentence into words. The key idea in such a decomposition is a substring.

A **substring** of a character string is any consecutive sequence of characters in the string. For example, "J", "ne D", and "Doe" are substrings of the character string "Jane Doe", but "JDoe" is not a substring. Every character string is regarded as a substring of itself. The string of length zero (the null string) is a substring of every string; it occurs between every pair of characters and at both the beginning and end of the string. The following table indicates all the substrings of the character string "then".

Length 0: " " (the null string)
Length 1: "t" "h" "e" "n"
Length 2: "th" "he" "en"

Length 3: "the" "hen"
Length 4: "then"

Referencing Substrings

There is a convenient way to refer to any contiguous subsequence of characters of a character string. This is done by writing after any character variable or array element two integer expressions that give the positions of the first and last characters in the substring. These two expressions are separated by a colon and enclosed in parentheses. The positions are numbered from 1 to *n*, where *n* is the length of the string. An example is `string(k:m)`, where the values of `k` and `m` are positive integers less than or equal to the length of `string` and `k` ≤ `m`. If `k` > `m`, the result is the null string. For example if `c` = "crunch",

`c(2:4)` is run
`c(1:6)` is crunch
`c(3:2)` is the null string
`c(2:7)` is illegal
`c(5:5)` is c

The last example illustrates how to refer to a single character of a string. The program `single_letters` tells the computer to print, one at a time, the characters of a string supplied as input.

```
program single_letters
!  Print individually the letters of an input string

   implicit none
   integer :: k
   character(len=10) :: string

   read "(a)", string
   print *, "Input data  string: ", string

   do k = 1, len(string)
      print *, "|", string(k : k), "|"
   end do

   print *, "====="
end program single_letters
```

```
 Input data  string: SHAZAM
 |S|
 |H|
 |A|
 |Z|
 |A|
 |M|
 | |
```

In a substring reference either of the lower or upper character positions may be omitted. In that case, 1 is used for the lower position and the length of the string is the upper position. For example

```
string = "abcde"
string(:3) is "abc"
string(3:) is "cde"
```

Reassigning the Value of a Substring

It is possible to reassign the value of a substring without affecting the rest of the string. For instance, the three lines

```
name = "John X. Public"
initial = "Q"
name(6:6) = initial
```

tell the computer to change the value of the variable `name` from "John X. Public" to "John Q. Public". Similarly the three lines

```
name = "John Xavier Public"
new_middle_name = "Quincy"
name(6:11) = new_middle_name
```

direct the computer to change the value of the variable `name` from "John Xavier Public" to "John Quincy Public".

In reassigning the value of a substring as in the above two examples, it is necessary that the length of the new substring value is exactly equal to the length of the old substring value. The following example shows how to accomplish something similar when the middle names are not the same length; in this case, `name` must be declared to be an allocatable character string.

```
name = "John Paul Public"
name = name(1:5) // "Peter" // name(10:16)
```

Finding the Position of One String in Another

There are numerous reasons for wanting to know if one string is contained as a substring in another. We might want to know if a particular letter is in a word or if a certain word is in a sentence. The built-in function `index` tells even more than that; it tells where to find the first instance of one character string as a substring of another. For example,

```
index("monkey", "on")
```

is 2 because the substring "on" begins at the second letter of the string "monkey" and

```
index("monkey", "key")
```

is 4 because the substring "key" begins at the fourth letter of "monkey".

If the string supplied as the second argument occurs more than once as a substring of the string supplied as the first argument, the function value is the location of the beginning of the leftmost occurrence, so that

```
index("banana", "ana")
```

is 2 even though characters 4–6 of "banana" are also "ana". If the second argument is not a substring of the first argument, rather than calling it an error and halting, a function value of zero is used as a signal. For example,

```
index("monkey", "off")
```

is 0. A program that calls the function `index` can test for the signal value zero if desired.

The intrinsic function `index` has an optional third argument `back` (backwards). When `back` is true, the search is for the rightmost occurrence of the substring. For example, `index("banana", "ana", .true.)` is 4, but `index("banana", "ana", .false.)` is 2. The default value for this argument is false.

Concatenation

The only built-in operation that can be performed on strings that produces a string result is **concatenation**. The concatenation of two strings is formed simply by placing one string after the other. The symbol for concatenation is two slashes (//). The program `plural` attempts to form the plural of given words by the method of putting the letter "s" at the end. Obviously, this program is not very useful as it stands, but it does illustrate the use of the concatenation operator.

```
program plural

   implicit none
   character(len=18) :: word
   integer :: ios

   do ! until out of words
      read (unit=*, fmt="(a)", iostat=ios) word
      if (ios < 0) exit  ! End of file
      print *, "Input data  word: ", trim(word)
      print *, "  Plural of word: ", trim(word) // "s"
   end do

end program plural
```

```
Input data  word: program
  Plural of word: programs
Input data  word: programmer
  Plural of word: programmers
Input data  word: matrix
  Plural of word: matrixs
Input data  word: computer
  Plural of word: computers
Input data  word: horses
  Plural of word: horsess
```

The `read` statement in the program `plural` needs both a format specification and an option `iostat` (input/output status) that sets the integer variable `ios` to a negative value when attempting to read beyond the end of the file. Thus, the long form (11.3) is required. However, we still want to use the default input unit, so we can write

```
read (unit=*, fmt="(a)", iostat=ios) word
```

The default input format is not used because we do not want to type quotes around the input string.

Exercises

1. What is the value of each of the following expressions?

```
len ("5 feet")
len ("alphabet")
len ("abcdefghijklmnopqrstuvwxyz")
len ("42")
```

2. List all the substrings of length 3 of the string "alphabet".

3. Write a program that reads a character string of maximum length 50 and prints all substrings of length 3. If you cannot think of anything better, use as input data

```
"These are the times that try our souls."
```

The output from this sample input should be

```
The
hes
ese
.
.
.
uls
ls.
```

4. Write a program that reads a character string of maximum length 10 and prints all of its substrings.

5. Write a program that sorts a list of at most 200 character strings. Each character string is at most 50 characters long and occupies the leftmost positions of one line in the input file. Use the end-of-file test to terminate reading of input data. You may be surprised at what happens if you accidentally type a blank in the leftmost column of one of the lines in the input file. Then again, after you think about it, you might not be.

6. A computer system maintains a list of valid passwords. Write a program that accepts an 18-character password and checks it against its list of valid passwords. The program should print "ok" if the password is in the list and "Try again" if it is not. Give the user two additional tries, replying with successively nastier messages each time the user fails to give the correct password. *Hint:* Keep a list of responses as well as a list of passwords. A sample execution might produce the following output:

```
  Welcome to the super special simulated system
  Enter your password:
bug free
  Try again
  Enter your password:
silicon
  Are you sure you have a password?
  Enter your password:
i love fortran
  ok
```

7. Read an allocatable character string as input and print it in reverse order. You should use a character-valued function `reverse(string)`. If the input is

```
until
```

the output should be

```
Input data  string:  until
litnu
```

8. Nicely displayed headings add impact to a document. Write a program to take a character string as input and print it surrounded by a border of exclamation marks. Again, ignore trailing blanks in the input. Leave one blank before the first character and after the last character in the display. If the input is

```
Payroll Report
```

the output should be

```
Input data  title:  Payroll Report

!!!!!!!!!!!!!!!!!!
! Payroll Report !
!!!!!!!!!!!!!!!!!!
```

9. Write a logical-valued function `fortran_name` that determines whether or not its character string argument is a legal Fortran name.

10. If a chess or checkers board is declared by

```
character(len=1), dimension(8, 8) :: board
```

the statement

```
board = "R"
```

assigns the color red to all 64 positions. Write a statement or statements that assigns "B" to the 32 black positions. Assume that `board(1, 1)` is to be red so that the board is as shown in Table 5-3.

Table 5-3 A chess or checkers board

R	B	R	B	R	B	R	B
B	R	B	R	B	R	B	R
R	B	R	B	R	B	R	B
B	R	B	R	B	R	B	R
R	B	R	B	R	B	R	B
B	R	B	R	B	R	B	R
R	B	R	B	R	B	R	B
B	R	B	R	B	R	B	R

11. Write a function `char_to_int` that accepts a character string and returns a vector of integers, one for each character in the string. The integer value should be 1 through 26, reflecting the position in the alphabet if the character is either an uppercase or lowercase letter; the value should be zero, otherwise. For example, `char_to_int("e")` = [5] and `char_to_int("a-z")` = [1 0 26].

12. Write a function `int_to_binary` that converts an integer to a character string that is the binary representation of the integer. Adjust the 1s and 0s in the right-hand portion of the string and pad the remainder of the string with blanks. The string should contain no insignificant zeros, except that the integer 0 should produce the string consisting of all blanks and one character "0". If the integer is negative, the first nonblank character should be a minus sign; if it is positive, the first nonblank character should be "1". Declare the function result to have length 5. If the declared length is not long enough to contain the result, it should consist of all asterisks.

`int_to_binary(5)` is *bb*101
`int_to_binary(0)` is *bbbb*0
`int_to_binary(-4)` is *b*–100

`int_to_binary(77)` is *****

You should write the function in such a way that changing the length of the function result to 16 requires no additional changes in the function.

13. Write and test a function `my_index` identical to the intrinsic function `index`. When `back` is true, the search proceeds backward from right to left.

5.2 Text Analysis

There are numerous reasons for examining text in minute detail, word by word and letter by letter. One of the reasons is to determine the authorship of an historical or literary work. Such quantities as the average length of a word or the frequency of usage of certain letters can be important clues. Computers have been useful in studying text from this viewpoint.

Blanking Out Punctuation

We start with some routines that perform simple text manipulation processes. The subroutine `blank_punct` (blank out punctuation) uses the substring value reassignment facility and the intrinsic function `index`. Keep in mind that a function value zero means the function `index` has determined that the second supplied argument is not a substring of the first supplied argument. The subroutine `blank_punct` regards any character besides a letter or a blank as a "punctuation mark" to be blanked out.

The program `test_bp` (test blank out punctuation) is intended to show how the subroutine `blank_punct` works. The dummy argument `text` is declared with length * because its length is not changed by the subroutine.

```
module blank_module

   implicit none
   public :: blank_punct

contains

subroutine blank_punct(text)
   ! Blank out punctuation
   ! Retain only letters and blanks

   character(len=*), intent(in out) :: text
   character(len=*), parameter :: letter_or_b =  &
         "ABCDEFGHIJKLMNOPQRSTUVWXYZ" //  &
         "abcdefghijklmnopqrstuvwxyz "
   integer :: i
```

```
   ! Replace any character that is not a blank
   ! or letter with a blank
   do i = 1, len_trim(text)
      if (index(letter_or_b, text(i:i)) == 0) then
         text(i:i) = " "
      end if
   end do
end subroutine blank_punct

end module blank_module

program test_bp

   use blank_module
   implicit none
   character(len=:), allocatable :: text
   text = "Suppress5$,superfluous*/3punctuation."
   call blank_punct(text)
   print *, text

end program test_bp
```

The result is

```
Suppress   superfluous   punctuation
```

A slightly different version of the subroutine `blank_punct` uses the `verify` built-in function. The `verify` function scans the first argument, checking that each character in the string is also in the string that is the second argument. If each character in the first argument is also in the second, the value of the function is 0. Otherwise, the value of the function is the character position of the leftmost character in the first argument that is not in the second argument. For example the value of `verify("banana", "nab")` is 0 and the value of `verify("banana", "ab")` is 3, the position in "banana" of the first letter that is neither "a" nor "b".

```
subroutine blank_punct(text)
! blank out punctuation
! retain only letters and blanks

   character(len=*), intent(in out) :: text
   character(len=*), parameter :: letter_or_b =  &
         "ABCDEFGHIJKLMNOPQRSTUVWXYZ" //  &
         "abcdefghijklmnopqrstuvwxyz "
   integer :: i

   ! Replace any character that is not a blank
   ! or letter with a blank
   do
      i = verify(text, letter_or_b)
      if (i == 0) exit
```

```
      text(i:i) = " "
   end do
end subroutine blank_punct
```

The subroutine blank_punct needs a character string parameter letter_or_b of length 53 that does not fit conveniently on one line. The concatenation operator is used to break the line.

Excising a Character from a String

When a character of a string is blanked out, as by the subroutine blank_punct, that character is replaced by a blank and the length of the character string remains unchanged. When a character is *excised* from a string, not only is the character removed, but also all of the characters to the right of the excised character are moved one position to the left. If name is fixed length, the character in position c of name can be excised by the statement

```
name(c:) = name(c+1:)
```

which adds a blank at the end of the string.

If name is allocatable, this does not work because the left side of the assignment is not a simple name, so the length is not changed and a blank is still added to the end of the string. If instead, the statement

```
name = name(:c-1) // name(c+1:)
```

is used, the blank character at position c is removed and the string is shortened by one character.

The subroutine compress_bb (compress double blanks) removes all double blanks from a string except those that occur at the right end. It is called by a program words that lists all the words in a string; the program words is discussed in the next subsection. Here the dummy argument text is declared to have length :.

```
subroutine compress_bb(text)
!  Removes double blanks
   character(len=:), intent(in out), allocatable :: text
   integer :: i

   do
      i = index(text, "  ")
      if (i == 0) exit
      text = text(:i-1) // text(i+1:)
   end do
end subroutine compress_bb
```

Listing All the Words

We now turn our attention to the problem of listing all the words in a text. For this purpose, the program words regards a substring as a word if and only if it consists entirely of letters and both the character immediately before it (if any) and the character immediately after it (if any) are not letters. The program does not consult a dictionary to see whether the word has been approved by a lexicographer.

For input, a fixed-length character string `text_input` is used because reading an allocatable string does not set its length. Then `text_input` (trimmed) is assigned to the allocatable string `text`.

```
program words

   use words_module
   implicit none
   character(len=200) :: text_input
   character(len=:), allocatable :: text
   integer :: end_of_word

   read "(a)", text_input
   text = trim(text_input)
   print *, "Input data  text: ", text

   ! Blanking out the punctuation,
   ! compressing the multiple blanks,
   ! and ensuring that the first character is a letter
   ! are pre-editing tasks to simplify the job.
   call blank_punct(text)
   call compress_bb(text)
   text = trim(adjustl(text)) // " "

   ! Print all the words.
   ! Each word is followed by exactly one blank.
   do ! until all words are printed
      if (len(text) == 0) exit
      end_of_word = index(text, " ") - 1
      print *, text(:end_of_word)

      ! Discard word just printed
      text = text(end_of_word+2:)
   end do

end program words
```

```
 Input data  text: Then, due to illness*, he resigned.
 Then
 due
 to
 illness
 he
 resigned
```

If the string supplied as input to the program `words` contains no letters, the pre-editing provides a string of all blanks to the **do** loop that prints all the words. The **do** loop exits correctly on the first iteration without printing any words because the trimmed length is zero. In the usual case, however, a word starts at position 1 of `text` and stops immediately before the first blank. The computer prints the word and discards it and the blank immediately following it, so that the next word to be printed begins at location 1 of the resulting character string.

Even after the subroutines `blank_punct` and `compress_bb` are called, it is possible that the first character of the string is a blank. Application of the built-in function `adjustl` shifts the string to the left to eliminate any leading blanks.

Average Word Length

To compute the average length of words in a given text, it is necessary to determine both the total number of letters in each word and the total number of words. The most direct way that comes to mind is used by the program `avg_word_len_1` (average word length, version 1).

```
program avg_word_len_1
   Initialize word count and letter count to zero
   Read text
   Start scan at leftmost character of the text
   Do until end of text is reached
      Locate the beginning and end of a word
      If no more words then exit the loop
      Increase the letter count
            by the number of letters in the word
      Increase the word count by 1
   Print "average word length = ", letter count / word count
end program avg_word_len_1
```

After reading in the text, the computer starts to look for the first word at the extreme left. Blanks, commas, and other nonletters are passed over to find the beginning of a word. Then letters are counted until the first nonletter is reached, such as a blank or punctuation mark, which signals the end of the word. These steps are repeated for each word in the text. Each time it locates a word, the computer increases the letter count by its length and the word count by one.

The refinement of `avg_word_len_1` is straightforward. It uses the intrinsic function `scan` that works like `verify`, except that it looks for the first occurrence of any character from a set of given characters, in this case the alphabetic characters.

```
program avg_word_len_1
!  Calculate the average word length of input text

   implicit none
   character(len=200) :: text
   integer :: word_begin, word_end
   integer :: word_count, letter_count
```

```
    character(len=*), parameter :: alphabet =  &
          "ABCDEFGHIJKLMNOPQRSTUVWXYZ" //  &
          "abcdefghijklmnopqrstuvwxyz"

    letter_count = 0
    word_count = 0
    read "(a)", text
    print *, "Input data  text: ", trim(text)

    do  ! until no more words
       word_begin = scan(text, alphabet)
       if (word_begin == 0) exit
       text = text(word_begin:)
       word_end = verify(text, alphabet) - 1
       if (word_end == -1) word_end = len(text)
       letter_count = letter_count + word_end
       word_count = word_count + 1
       text = text(word_end+2:)
    end do

    print *, "Average word length =",  &
          real(letter_count) / word_count

end program avg_word_len_1
```

```
 Input data  text: Never mind the whys and wherefores.
 Average word length =    4.8333335

 Input data  text: I computed the average word length.
 Average word length =    4.8333335
```

The sample execution output of the program avg_word_len_1 might suggest that to use average word length as a test for authorship, one should have a fairly large sample of text.

Modification for a Large Quantity of Text

If the amount of text is very large, the computer might not have enough memory to hold it all at one time. Thus it may be desirable to modify the program avg_word_len_1 so that it reads the text one line at a time, rather than all at once. The program avg_word_len_2 incorporates such a modification. Much of the main program avg_word_len_1 is put into the subroutine one_line (process one line).

```
module word_length_2_module

   implicit none
   public :: one_line
```

```
   character(len=*), parameter, private :: alphabet = &
         "ABCDEFGHIJKLMNOPQRSTUVWXYZ" // &
         "abcdefghijklmnopqrstuvwxyz"
   character(len=200), public :: text
   integer, public :: word_count, letter_count

contains

subroutine one_line()
!  Accumulate statistics on one line of input text.
   integer :: word_begin, word_end

   do  ! until no more words
      word_begin = scan(text, alphabet)
      if (word_begin == 0) exit
      text = text(word_begin:)
      word_end = verify(text, alphabet) - 1
      if (word_end == -1) then
         word_end = len(text)
      end if
      letter_count = letter_count + word_end
      word_count = word_count + 1
      text = text(word_end+2:)
   end do

end subroutine one_line

end module word_length_2_module

program avg_word_len_2
!  Calculate the average word length of input text.
!  Text may have many lines, terminated by end of file.

   use word_length_2_module
   implicit none
   integer :: ios

   letter_count = 0
   word_count = 0

   do  ! until no more lines of text
      read (unit=*, fmt="(a)", iostat=ios) text
      if (ios < 0) exit
      print *, "Input data  text: ", trim(text)
      call one_line()
   end do

   print *, "Average word length =",  &
         real(letter_count) / word_count
end program avg_word_len_2
```

```
Input data  text: One of the more important uses
Input data  text: of the character manipulation
Input data  text: capability of computers is
Input data  text: in the analysis of text.
Average word length =    4.8947368
```

Frequency of Occurrence of Letters

There are two basic ways to count the number of occurrences of each letter of the alphabet in a given text. Both ways use 27 counters, one for each letter of the alphabet and one to count all the other characters.

One way to tabulate letter frequencies in a line of text is first to scan it for all occurrences of the letter "a", then to scan it for all occurrences of the letter "b", and so on through the alphabet. This requires 26 scans of the whole line. This method is embodied in the pseudocode program `letter_count_1`.

```
program letter_count_1
   Initialize
   do
      Read line of text
      If no more text, exit loop
      do letter = "a", "z"
         Scan line of text, counting occurrences of that letter
               (either uppercase or lowercase)
         Calculate the number of nonletters and
               increment nonletter total
      end do
   end do
   Print the counts
end program letter_count_1
```

The second way to count letter frequencies in a line of text is to begin with the first symbol of the text, to decide which of the 27 counters to increment, to continue with the second letter of the line of text, to see which counter to increment this time, and so on through the text. This second way is implemented by the pseudocode program `letter_count_2`.

```
program letter_count_2
   Initialize
   do
      Read a line of text
      If no more text, exit loop
      Do for each character in the line of text
         If the character is a letter then
            Increment the count for that letter
```

```
            else
               Increment the nonletter count
            end if
         end do
      end do
      Print the counts
   end program letter_count_2
```

By the method of the program `letter_count_1`, the text must be scanned completely for each letter of the alphabet. By the method of the program `letter_count_2`, the text is scanned just once. Thus, the second program executes considerably faster than the first one and so only the program `letter_count_2` is supplied below.

In Fortran, the subscripts of the array of counters cannot be "a", "b", etc. A subscript must be an integer. Therefore, subscripts 1 through 26 are used to count the number of occurrences of each letter of the alphabet and subscript 0 is used to count the characters that are not letters.

```
module letter_count_2_module

   implicit none
   public :: count_letters, print_counts

!  Variables:
!     tally(0) = count of nonletters
!     tally(1) - tally(26) = counts of A/a - Z/z

   character(len=*), parameter, private :: alphabet =  &
         "ABCDEFGHIJKLMNOPQRSTUVWXYZ" //  &
         "abcdefghijklmnopqrstuvwxyz"
   character(len=200), public :: text
   integer, dimension(0:26), public :: tally

contains

subroutine count_letters()
!  Count letters in one line of text
   integer :: i, letter

   do i = 1, len_trim(text)
      letter = index(alphabet, text(i:i))
      if (letter > 26) letter = letter - 26
      tally(letter) = tally(letter) + 1
   end do
end subroutine count_letters

subroutine print_counts()
!  Print the frequency counts

   integer :: letter
```

```
   print *
   print "(2a10)", "Letter", "Frequency"
   do letter = 1, 26
      print "(a10, i10)",  &
            alphabet(letter:letter), tally(letter)
   end do
   print "(a10, i10)", "Other", tally(0)
end subroutine print_counts

end module letter_count_2_module

program letter_count_2
!  Count frequency of occurrence in a text
!  of each letter of the alphabet

   use letter_count_2_module
   implicit none
   integer :: ios

   tally = 0  ! Set entire array to zero

   do  ! until no more lines in file
      read (unit=*, fmt="(a)", iostat=ios) text
      if (ios < 0) exit
      print *, "Input data  text: ", trim(text)
      call count_letters()
   end do

   call print_counts()

end program letter_count_2
```

```
 Input data  text: One of the important text analysis
 Input data  text: techniques (to determine authorship)
 Input data  text: is to make a frequency count of
 Input data  text: letters in the text.

    Letter Frequency
         A         6
         B         0
         C         3
         D         1
         E        15
         F         3
         G         0
         H         5
         I         7
         J         0
         K         1
```

```
        L      2
        M      3
        N      8
        O      8
        P      2
        Q      2
        R      5
        S      6
        T     16
        U      4
        V      0
        W      0
        X      2
        Y      2
        Z      0
    Other     20
```

This is a good example to illustrate how and where different variables and parameters are declared. The variable `count` is a public variable in the module because it is used in the program `letter_count_2`. The parameter `alphabet` and the variable `text` are private because they are not needed in the program `letter_count_2`, but are above the `contains` statement because they are used in both of the module procedures. The remaining variables are declared within the subroutines because they are needed in only one subroutine.

Palindromes

Another aspect of text analysis is searching for patterns. Perhaps the text repeats itself occasionally, or perhaps the lengths of the words form an interesting sequence of numbers. One pattern for which we search here is called a "palindrome", which means that the text reads the same from right to left as from left to right. The word "radar" is a palindrome, for example. Liberal palindromers customarily relax the rules so that punctuation, spacing, and capitalization are ignored. To liberal palindromers, the names "Eve", "Hannah", and "Otto" are all palindromes, as is the sentence

> "Able was I ere I saw Elba."

something Napoleon might have said, except that he preferred speaking French.

The program `palindrome` satisfies the most conservative palindromers. As the two sample runs show, it accepts the string

> "NAT SAW I WAS TAN"

as a palindrome, but it rejects the string

> "MADAM I'M ADAM"

It is straightforward to modify the program palindrome to apply a more liberal test for palindromes; simply preprocess the text as in the program words above. The subroutine blank_punct converts all nonletters to blanks, the subroutine compress_bb can be modified to excise all blanks, and a subroutine fold_cases can be written to change all lowercase letters to uppercase.

```
module c_or_blank_module

   implicit none
   public :: c_or_blank

contains

function c_or_blank(c) result(c_or_blank_result)
!  Tests if c is blank
!  Returns "blank" if it is
!  Returns c otherwise
   character(len=*), intent(in) :: c
   character(len=5) :: c_or_blank_result

   if (c == " ") then
      c_or_blank_result = "blank"
   else
      c_or_blank_result = c
   end if
end function c_or_blank

end module c_or_blank_module

program palindrome
!  Tests for a palindrome

   use c_or_blank_module
   implicit none
   character(len=200) :: text
   integer :: l, left, right
   logical :: match

   read "(a)", text
   print *, "Input data  text: ", trim(text)

   right = len_trim(text)
   match = .true.
   do l = 1, right / 2
      if (text(l:l) /= text(right:right)) then
         left = l
         match = .false.
         exit
```

```
         else
            right = right - 1
         end if
      end do

      if (match) then
         print *, "Palindrome"
      else
         print *, "Not a palindrome"
         print*, "Character",left,"from the left is ", &
               c_or_blank(text(left:left))
         print*, "Character",left,"from the right is ", &
               c_or_blank(text(right:right))
      end if

   end program palindrome
```

```
Input data  text: NAT SAW I WAS TAN
Palindrome

Input data  text: MADAM I'M ADAM
Not a palindrome
Character 5 from the left is M
Character 5 from the right is blank
```

Exercises

1. Write a more efficient version of `compress_bb` described in 5.1. When a double blank is found remove all of the consecutive blanks with one assignment statement.

2. Calculate the ratio of letters in the first half of the alphabet to letters in the second half of the alphabet in an input text.

3. An alliteration is a sequence of words all starting with the same letter. Write a program `alliteration` that counts the most consecutive words in an input text starting with the letter "P" or "p". For the sample input data

   ```
   In his popular paperback, "Party Pastimes People
   Prefer", prominent polo player Paul Perkins
   presents pleasing palindromes.
   ```

   the output should be

   ```
   14
   ```

5.3 Case Study: Expression Evaluation

In 3.16 it was mentioned that it is possible for recursive procedures to call each other. This is illustrated in this section with an example that also gives a little insight into how computer programs are processed, producing the answers that we expect to see when a program is run.

In this book, the syntax or form of Fortran statements is given by a very informal description. The following definitions use a more formal notation to describe a small part of Fortran, namely, a class of arithmetic expressions involving only nonnegative integer constants, addition (+), multiplication (*), and parentheses. A more complete description of this notation can be found in Appendix B.

The first thing to do is describe what a number is.

number **is** *digit*
or *digit number*

This says that a number is either a single digit or a digit followed by another (shorter) number. It is a recursive definition because the definition of *number* involves *number* as part of the second option. As with any recursive definition or program, there must be a way to terminate the recursion; in this case, a *number* must eventually be just a *digit*, the first choice for *number*. A digit is a single character 0, 1, ..., or 9. This is a situation in which the recursion is not very essential and a number can be described more simply as a sequence of one or more digits, but this provides a very simple example of the definitions of other syntactic objects that are a little more complicated.

The fundamental building block of an expression is called a *primary*. Primaries are the basic components out of which expressions are built; they are treated as operands and combined using arithmetic operators. In our case it is either a number or any other expression enclosed in parentheses.

primary **is** *number*
or (*expression*)

The next rule indicates how to build expressions using just primaries and multiplication to form what are called *terms*. A term is a sequence of primaries separated by the multiplication symbol (*). It can also be described recursively with the rule

term **is** *primary*
or *primary* * *term*

This description says that a *term* is either a *primary* by itself or a *primary* followed by the multiplication symbol and another (simpler) term. Examples of terms are

```
64
111*2222
397*43*(2899*64352)
```

In the last case one of the primaries is `(2899*64352)`, which is an example of an expression enclosed in parentheses.

The description of an expression is similar to that of a term. An expression is a sequence of terms separated by the plus (+) operator and can be described recursively in our notation by

expression **is** *term*
or *term* + *expression*

Examples of expressions are all of the example terms given in the previous list and the following as well.

```
111+2222
111+2222*33
(111+2222)*33
```

The last two are seen to be expressions in slightly different ways. For `111+2222*33`, `111` is a term and `2222*33` is an expression, because it is a term consisting of two primaries separated by *. However, `(111+2222)*33` consists of an expression in parentheses followed by * and the number 33. This illustrates that the syntax rules indicate how the expression is to be broken down into components, which, in turn, indicates how the value of the expression will be computed.

It is now possible to see how intertwined these definitions are. We started with the definition of a primary that involved an expression. But the definition of expression involves the definition of term, which involves the definition of primary!

It is possible to construct a program that determines if a string of characters is a legal expression as defined above. This program can be implemented using the recursive definitions directly or it can take advantage of the *tail recursion* in the definitions of *number, term,* and *expression* in order to be a little more efficient. However, it is not as easy to see how to handle the second alternative in the definition of primary without a recursive call to determine if it is an expression in parentheses.

It is interesting that it is possible to write a program that is not much more complicated than one that just tests the legality of an expression, but that computes the *value* of any legal expression. Giving the rules that determine the value of each expression specifies the *semantics* or meaning of the expression. It is easy to transform the rules given above into rules that compute the value of any expression.

A primary is either a number or an expression in parentheses; the value of a primary is either the value of the number or the value of the expression in parentheses. This sounds so simple that it may seem like it does not even say anything, but it does give the value of any primary in terms of its components. By the way, we will assume that the value of a number is "obvious", although it is not hard to define the value of a number in terms of its digits.

The description of the value of a term and an expression are very similar. If a term is a primary, its value is the value of the primary, which is defined in the previous paragraph. If it is *primary * term,* its value is the product of the value of the primary and the term. Similarly, the value of expression is either the value of a term or the sum of the values of a term and another expression.

We can now begin to write some of the functions that will return the value of the various kinds of expressions. Blanks are not permitted or are removed by preprocessing. Taking them in the same order as before, **primary** is a function that computes the value of a primary. We agree to return the value –1 if the string is not a legal primary. This works because only nonnegative integer constants are allowed and there is no subtraction operator.

```
recursive function primary(string) result(primary_result)

   character(len=*), intent(in) :: string
   integer :: primary_result
   integer :: ls

   ! See if it is a number
   primary_result = number(string)
   ls = len(string)
   ! If not, see if it is an expression in parens
   if (primary_result < 0 .and. ls > 0) then
      if (string(1:1) == "(" .and.  &
          string(ls:ls) == ")") then
         primary_result = expression(string(2:ls-1))
      end if
   end if

end function primary
```

The first executable statement evaluates the primary as if it were a number. If it is a number, the value is not –1 and its value is also the value of the primary. If the value is –1, the other option is that the primary is an expression enclosed in parentheses. The first and last characters are checked—if they are left and right parentheses, respectively, the expression between is evaluated and is used as the value of the primary.

To check that a string is a number, which must be a string of one or more digits, the **verify** function is used. It returns 0 if all the characters are digits. Also, if the length of the string is greater than zero, an internal **read** statement (11.3) is used to convert the string of digits to an integer value.

```
function number(string) result(number_result)

   character(len=*), intent(in) :: string
   integer :: number_result

   ! Check that it is one or more digits
   if (len(string) > 0 .and.  &
         verify(string, "0123456789") == 0) then
      read (unit=string, fmt=*) number_result
   else
      number_result = -1
   end if

end function number
```

The function term that returns the value of a string if it is a term and –1 otherwise first checks to see if the string is a primary. If it is, the value of the primary is the value of the term. If it is not a primary, it must be a primary followed by * followed by another term.

```
recursive function term(string) result(term_result)

   character(len=*), intent(in) :: string
   integer :: term_result
   integer :: op

   ! Check if it is a primary
   term_result = primary(string)
   if (term_result < 0) then
      ! If not a primary,
      ! find the first * outside parens
      op = position(string, "*")
      if (op > 0) then
         term_result = &
            combine(primary(string (:op-1)), &
                       term(string (op+1:)), "*")
      else
         term_result = -1
      end if
   end if

end function term
```

We have made the function a bit more efficient by realizing that a primary cannot contain a multiplication sign unless it is inside parentheses. So we look for the leftmost multiplication sign that is not enclosed in parentheses. This is done by scanning the string, counting a left parenthesis as +1 and a right parenthesis as –1 and finding the first * at a place where the count is zero (and hence the parentheses to the left are balanced). The function position does this and returns 0 if it does not find such a multiplication symbol.

```
function position(string, op_symbol) result(position_result)

   character(len=*), intent(in) :: string, op_symbol
   integer :: position_result
   integer :: p, paren_count

   position_result = 0
   paren_count = 0
   do p = 1, len(string)
      if (string(p:p) == "(") then
         paren_count = paren_count + 1
      else if (string(p:p) == ")") then
         paren_count = paren_count - 1
```

```
      else if (string(p:p) == op_symbol .and. paren_count == 0) then
         position_result = p
         exit
      end if
   end do

end function position
```

It is interesting to note that a case construct cannot be used to select which operation to perform because the items in parentheses in each case statement must be *constants*; an item cannot be the character string that is the dummy argument op_symbol.

If position is positive, the function term treats the characters to the left of the * as a primary and the characters to the right as another term, getting their values and multiplying them together if neither is –1. The function combine is used to multiply two values together, except that it returns –1 if either argument is –1. It is also used to add two values, so it takes a third argument that indicates which operation to perform.

```
function combine(x, y, op_symbol) result(combine_result)

   integer, intent(in) :: x, y
   character(len=*), intent(in) :: op_symbol
   integer :: combine_result
   if (x < 0 .or. y < 0) then
      combine_result = -1
   else
      select case (op_symbol)
         case ("+")
            combine_result = x + y
         case ("*")
            combine_result = x * y
         case default
            combine_result = -1
      end select
   end if

end function combine
```

The function expression is very similar to term. The names of the functions called are changed, and the operator passed to combine is + instead of *.

All of these functions are put in a module. This example does not seem to be improved by using allocatable strings; in almost all cases, a string is a dummy argument, which is declared to have length *, so a dynamically changing length is not needed.

```
module expression_module

   implicit none
   public :: expression, term, primary, number, position, combine

contains
```

```
recursive function expression (string) . . .

recursive function term (string) . . .

recursive function primary (string) . . .

function number (string) . . .

function position (string, op_symbol) . . .

function combine (x, y, op_symbol) . . .

end module expression_module

program expression_evaluation

   use expression_module
   implicit none
   character(len=100) :: line
   integer :: status, value

   do
      read (unit=*, fmt="(a)", iostat=status) line
      if (status < 0) exit
      print *
      print *, "Input data  line:  ", trim(line)
      value = expression(trim(line))
      print *, "The value of the expression is: ", value
   end do

end program expression_evaluation

Input data  line:  (443+29)(38+754)
 The value of the expression is:  -1

 Input data  line:  89+23*4
 The value of the expression is:  181

 Input data  line:  (((((((555)))))))
 The value of the expression is:  555

 Input data  line:  64+23*(5388+39)*(54*22+3302*2)
 The value of the expression is:  972605296
```

Exercises

1. Give a recursive definition of the value of a *number* that uses the value of a digit.

2. Extend the `expression_evaluation` program to allow negative constants, subtraction, and division. The result should be `-huge(0)` if the string is not a legal expression.

6 Structures and Derived Types

Fortran arrays allow data to be grouped, but only if all items have the same data type. It is often useful to use a structure, which is a compound object consisting of values that may be of different data types. Derived types are used to define the form of structures. It is possible to define new operations and functions on defined types, creating abstract data types. Derived types and their operations are defined in a module, making them globally available to many programs.

An interesting kind of structure is a recursive data structure, which can be built and manipulated using pointers. Examples of these structures are found in the form of linked lists (10.3), trees (10.4), and queues (12.4).

6.1 Structures

A **structure** is a collection of values, not necessarily of the same type. The objects that make up a structure are called its **components**. The components of a structure are identified by Fortran names, whereas the elements of an array are identified by numerical subscripts.

A good example of the use of a structure might be provided by a simple text editor, such as one supplied with many programming language systems. Each line in a program consists of a line number and one or more statements. When the editor is running, the program being edited could be represented in the editing program as two arrays, one to hold line numbers and one to hold the text of each line. Perhaps a better way to do this is to have a single object called `line` consisting of two components, an integer `line_number` and a character string `statement`. The entire program would then be an array of these structures, one for each line.

The components of a structure may be arrays or other structures. The elements of an array may be a structure. The elements of an array may not be arrays, but this functionality can be achieved with an array whose elements are structures whose only component is an array or by a higher dimensional (rank) array.

To give a slightly more complicated example, suppose we wish to store in our computer the contents of our little black book that contains names, addresses, phone numbers, and perhaps some remarks about each person in the book. In this case, each entry in the book can be treated as a structure containing four components: name, address, phone number, and remarks. The diagram in Figure 6-1 represents the organization of this information.

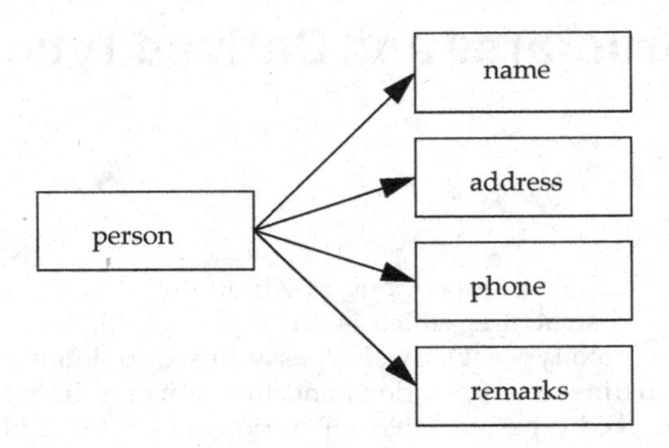

Figure 6-1 Diagram of the structure person

The name of the structure is `person`, and it has four components: `name`, `address`, `phone`, and `remarks`. Sometimes one or more components might be broken down into lower-level components. For instance, if the owner of the black book wanted to contact every acquaintance in a particular city, it would be helpful to have the component `address` itself be a structure with components `number`, `street`, `city`, `state`, and postal `zip_code`. With this organization of the data, it would be possible to have a computer program scan the entries for city and state without having to look at the street address or zip code. For similar reasons, it might be convenient to subdivide each telephone number into a three-digit area code and a seven-digit local number, assuming all of the numbers are in North America. This more refined data organization is represented schematically by the structure in Figure 6-2.

6.2 Derived Types

As was discussed in 1.2, there are five intrinsic Fortran data types: integer, real, complex, logical, and character. A programmer may **define** a new data type, called a **derived type**. In Fortran, a derived type can be used only to define a structure. Conversely, a structure can occur in a program only as a value of some derived type.

A **type definition** begins with the keyword `type`, followed by either the `private` or `public` accessibility attribute (assuming it is in a module), followed by two colons (`::`) and the name of the type being defined. The components of the type are given in the form of ordinary type declarations. The type definition ends with the keywords `end type`, followed by the name of the type being defined.

Style note: All type definitions should be put above the `contains` statement in a module. In other words, a type definition should not appear in a main program or a procedure. Each should be given the `public` or `private` attribute.

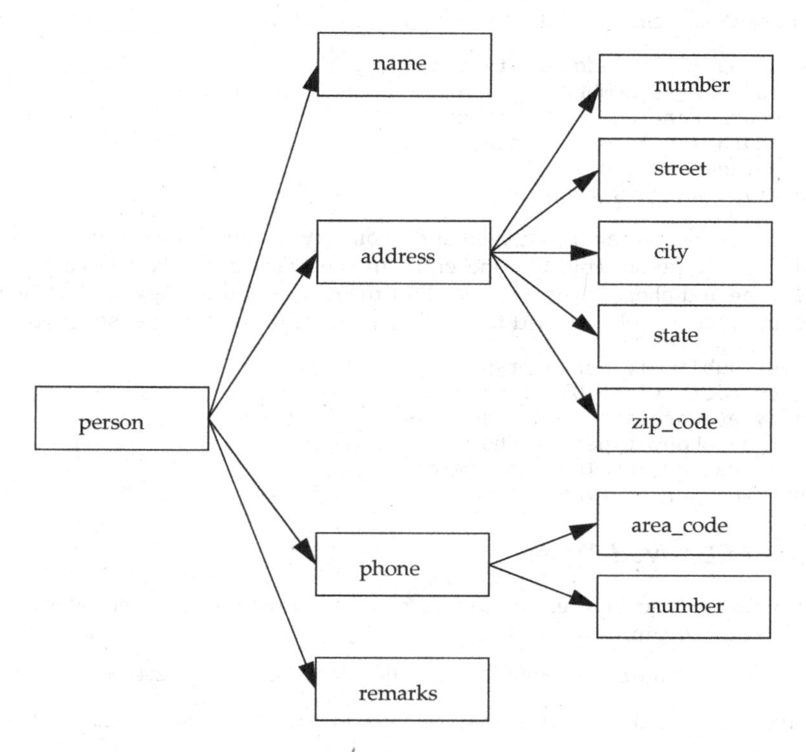

Figure 6-2 A refined structure person

Let us start first with the program editor example, for which each line of the program consists of a line number and some text. A definition of a type that would be useful in this example is

```
type, public :: line
   integer :: line_number
   character(len=line_length) :: text
end type line
```

where `line_length` is an integer parameter (named constant).

Let us next return to the example of the little black book. To define the type `phone_type` in that example, `area_code` and `number` are each declared to be integer components:

```
type, public :: phone_type
   integer :: area_code, number
end type phone_type
```

The definition of the type address_type is a little more complicated because some of the components are character strings and some are integers:

```
type, public :: address_type
   integer :: number
   character(len=30) :: street, city
   character(len=2) :: state
   integer :: zip_code
end type address_type
```

Now that the types address_type and phone_type have been defined, it is possible to define a type suitable for one entry in the black book. Note that the names address_type and phone_type were used for the names of the types, so that the names address and phone could be used for the components of the type person_type.

```
type, public :: person_type
   character(len=40) :: name
   type(address_type) :: address
   type(phone_type) :: phone
   character(len=100) :: remarks
end type person_type
```

Features of Derived Types

Derived types have several features, some of which must be described later in combination with other features of Fortran.

- Components of a derived type may be allocatable (4.1, 10.4, 12.4) or a pointer (10).
- Components of a derived type may be a procedure pointer (10.1, 12.3, 13.15, 13.17).
- A derived type may have a final procedure (10.3), which is executed whenever a variable of the type is destroyed.
- A derived type may have a type-bound procedure associated with it.
- A derived type may have parameters that are specified when a structure of the type is declared.

The last two features are introduced in the next two subsections.

Type-Bound Procedures

A derived-type declaration may have a contains statement followed by a **type-bound procedure**. The procedure is intended to process objects of the type or perform other functions related to the type. In many cases, a type-bound procedure processes an object of the type, so the default is that such an object is passed to the procedure unless the nopass attribute (12.3) is given to the procedure. An example using pass occurs in 12.4.

In the following simple example, the subroutine open_files might open some files used in processing objects of type t. The type-bound procedure open_files with the nopass attribute is named after the contains statement in the type definition and the procedure itself occurs after the contains statement of the module. Note that in the main program, the procedure is called by writing tt%open_files.

```
module m

   implicit none
   private

   type, public :: t
      real:: x=3.3
   contains
      procedure, nopass :: open_files
   end type

contains
   subroutine open_files()
      print *, "Opening file ..."
      ! open ( . . .
   end subroutine open_files

end module


program p

   use m
   implicit none

   type(t) :: tt
   call tt%open_files()


end program p
```

Parameterized Derived Types

All of the intrinsic types have a kind type parameter that allows the Fortran programmer to select a representation for the values of the type. In addition, the character type has a length type parameter. Similarly, a derived type may have type parameters, as specified by the programmer. There are two sorts of type parameters: kind and length. The basic difference between them is that the kind must always be given as an expression that can be evaluated at compiler time—hence, must consist essentially of constants, whereas a length parameter can change at runtime, just like character length or array bounds.

An example of a derived-type definition for a type with both kind and length parameters follows.

```
type, public :: matrix(rows, cols, k)
   integer, len :: rows, cols
   integer, kind :: k = kind(0.0)
   real(kind = k), dimension(rows, cols) :: values
end type matrix
```

The derived type `matrix` has three type parameters, listed in parentheses following the type name in the first line of the type definition. The names of the type parameters are `rows`, `cols`, and `k`. The type of each of these parameters, as well as an indication as to whether they are kind or length type parameters, must be given in declarations in the derived-type definition. These declarations are the second and third lines of the definition. The parameter `k` has a default value, so may be omitted when the type is used.

The next line lists the only component of the derived type: `values`, which is an array of real kind `k` with the specified number of rows and columns (`cols`). An example of a declaration of a variable of type `matrix` is given in 6.3.

Exercises

1. Design a data structure suitable for information on a college student to be used by the college registrar. Write the type definitions needed for this data structure.

2. Assuming that airlines accept reservations for flights up to one year in advance, design a data structure suitable for storing information associated with each reservation. Write the type definitions needed for this data structure.

3. Design a data structure suitable to hold information on each flight to be made by an airline during the next year. Write the type definitions needed for this data structure.

4. Design a data structure suitable for a bank to keep the information on a checking account. Write the type definitions needed for this data structure.

5. Add a fourth type parameter `rck` to the derived type `matrix`, given above. `rck` should be an integer kind type parameter that indicates the kind of the type parameters `rows` and `cols`.

6.3 Declaring and Using Structures

Given the type definition for `line` in 6.2 that can be used to hold a line number and one line of a program, a variable `new_line` that could be used to represent one line of the program can then be declared by

```
type(line) :: new_line
```

As shown in this example, a variable is declared to be a derived type with a declaration that is similar to the declaration of a variable of intrinsic type, except that the name of the intrinsic type is replaced by the keyword `type` and the name of the type in parentheses. Note that in a type *definition*, the name of the type is not enclosed in parentheses, but in a type *declaration*, the name of the type is enclosed in parentheses.

The entire program to be edited could be represented by a single variable declared to be an array of values of type `line`:

```
type(line), dimension(max_lines) :: program
```

With this declaration, some parts of the editor are a little easier to write and read because any operations that must be done to both a line number and the text can be expressed as a single operation on a line. For example, if two arrays were used, the `print` statement of a subroutine `list_program` might have been written

```
print "(i5, tr1, a)", line_number(n), text(n)
```

It can now be written

```
print "(i5, tr1, a)", program(n)
```

where the `tr` (tab right) edit descriptor (11.8) indicates a number of print positions to skip.

To use the type declarations for the address book, `joan` can be declared to be type `person_type` with the statement

```
type(person_type) :: joan
```

and the little black book can be declared to be an array of type `person_type`:

```
type(person_type), dimension(1000) :: black_book
```

Of course, any program or module that is to contain a derived-type *declaration* must use the module containing the derived-type *definition* (or be in the module containing the definition).

Variables can be declared to be type `matrix` (6.2) with a declaration like the following.

```
type (matrix (rows=40, cols=50) :: m
type (matrix (55, 65, kind=double)) :: b, c
```

These indicate that `m` is type `matrix` whose component `values` is a 40×50 array of default reals and that `b` and `c` are type `matrix` and have a component that is a 55×65 array of reals of kind `double`. In this situation, `double` must be an integer parameter defined somewhere previously (perhaps in a module that is used by the program).

An allocatable matrix with values of type default real can be declared using colons for the length parameters:

```
type (matrix(rows=:, cols=:, k=kind(0.0)), allocatable :: A
```

Referencing Structure Components

A component of a structure is referenced by writing the name of the structure followed by a percent sign (%) and then the name of the component. Suppose `joan` is a variable declared to be type `person_type` as shown above. Then Joan's address is referenced by the expression

```
joan % address
```

Style note: Blanks are permitted, but not required, around the percent sign in a structure component reference. We often use the blanks because it improves readability.

The object `joan % address` is itself a structure. If it is desired to refer to one of the components of this structure, another percent symbol is used. For example, the state Joan lives in is

```
joan % address % state
```

and her area code is

```
joan % phone % area_code
```

To see how structures can be used in a program, suppose the contents of the little black book are stored as the value of the variable `black_book` declared above, and suppose we want a subroutine that will print out the names of all persons who live in a given postal zip code. The subroutine simply goes through the entire contents of the black book, one entry at a time, and prints out the name of any person with the appropriate zip code.

Style note: If you use names followed by the suffix `_type` to name derived types, the same name without the suffix is available for variables and structure components of that type. For example, the component `name` can be type `name_type` and `address` can be `address_type`. This convention is used frequently in this book, but not always.

Default Initialization

There is no way in Fortran to indicate that (for example) every real variable should be initialized to zero (although some compilers have a compile option to do this). Something like this can be done with derived types, however. **Default initialization** indicates that certain components of a derived type are to be set to a specified value each time an object of that type is created by a declaration or allocation. In the following example, the `x` and `y` components of structures of type `point` are always initialized to 1.1 and 2.2, respectively. (To create a simple example, we violate the style rule that type definitions should be in a module.)

```
program default_initialization

   type :: point
      real :: x = 1.1, y = 2.2
   end type point

   type(point) :: p
   print *, p % x, p % y

end program default_initialization
```

The first executable statement of this program prints the two components of p. Except for a variable of derived type that is initialized or default initialized, this is not legal because it would not have a predictable value.

There is an example of default initialization in the linked-list program in 10.3.

Structure Constructors

Each derived-type definition creates a **structure constructor**, whose name is the same as that of the derived type. For example, if you define a type named boa, you have a boa constructor. This constructor may be used much like a function to create a structure of the named type. The arguments are values to be placed in the individual components of the structure. For example, using the type phone_type in 6.2, an area code and telephone number may be assigned with the statement

```
joan % phone = phone_type(505, 2750800)
```

It is not necessary that the function arguments be constants. If john also is type person_type and john % address has been given a value, the variable joan of type person_type can be assigned a value with the statement

```
joan = person_type ("Joan Doe", john % address,  &
       phone_type (505, fax_number - 1),  &
       "Same address as husband John")
```

A structure constructor looks a lot like a function call. One of the properties it shares with a function call is that the arguments may be specified by a keyword. Thus, the assignment to joan % phone above may be written

```
joan % phone = phone_type(area_code=505, number=2750800)
```

Also, a default initialized component may be omitted in a structure constructor.

```
type :: point
   real :: x=0.0, y=0.0
end type point

type (point) :: p
   . . .
p = point(y = 4.4)
```

A constructor for a parameterized derived type must include values for the parameters that do not have default values. For example, using the declarations from above

```
m = matrix(40, 50)(t1)
A = matrix(55, 65, double)(t2)
```

where t1 and t2 are real arrays of the appropriate shape.

Exercises

1. Write a program that builds a small database of friends' addresses and phone numbers by using the type definitions and declarations in this chapter. The program should prompt the user for information about each entry, keep the entries in an array, and write the whole database to a file when the program is terminated. The following statements from Chapter 11 should be of interest:

```
open (unit=19, file="pfile", status="replace", action="write")

write (unit=19, fmt="(...)") black_book(entry_number)

close (unit=19, status="keep")
```

   The open statement establishes that any operations using input/output unit 19 will refer to the file named "pfile". "replace" indicates that a new file is to be built, replacing any existing file with the same name. "keep" indicates that the file is to be kept when the program that builds the database stops.

2. Write a program that finds entries in the database created in Exercise 1 based on information provided by the user of the program. The statement

```
open (unit=19, file="pfile", status="old", &
      position="rewind", action="read")
```

   can be used to connect unit 19 to file "pfile" created by the program in the previous exercise. "old" indicates that the file is already there; "rewind" says to position the file at its beginning; and "read" indicates that the file will only be read and nothing will be written to it.

3. Write a function matrix_add(a, b) that produces the result of adding two objects of type matrix, with kind parameter default real, as defined in this section. If a and b are not conformable, the result should be a 0×0 matrix. Test the function.

7 IEEE Arithmetic and Exceptions

There is a standard that describes how real values should be stored in a computer. This standard also indicates the results when two values are added or multiplied. This seems obvious, but remember that most real values cannot be represented exactly in a computer, so the result of adding two numbers might be one of several possibilities that are approximations to the sum; which possibility is selected depends upon the rounding mode.

The use of the standard by most modern computers means that computations should be much more similar when performed on different computer systems.

This standard is often called the IEEE arithmetic standard, because it was first published as IEEE 754 (IEEE is the Institute of Electrical and Electronics Engineers). It is currently published by the International Standards Organization as IEC 60559 (1989-01), *Binary floating-point arithmetic for microprocessor systems*.

7.1 Numerical Representations

All values are stored in a computer as a finite collection of binary digits (bits). This means that most real values cannot be represented exactly, as discussed briefly in 1.7. It also means there is a limit to the size of numbers, both real and integer, that can be represented as values of a Fortran variable.

Representations of Integers

Integers (whole numbers) are represented as a string of bits. The number of bits is usually a power of two, such as 4, 8, 16, 32, or 64. For most Fortran systems, default integers are stored using 32 bits. The left bit is usually reserved for the sign: 0 for positive and 1 for negative. Using a 4-bit representation as an example, the integers 0, 1, 2, and 3 would be represented as 0000, 0001, 0010, and 0011, respectively. To understand how negative numbers are represented, think of a car odometer going backward. If the odometer is 0000 and the representation is binary, going backward one yields 1111, which is the representation of –1. 1110 is the representation of –2, and so forth. Other schemes may be used to represent negative integers, but this one is the most common.

With 32 bits, the largest positive number that can be represented is a 0 followed by thirty-one 1s, which represents the integer $2^{31} - 1$ or 2,147,483,647. This is the value of the intrinsic function `huge(0)`.

Representations of Reals

To understand how real values are stored, think of them as being written in exponential form, except that the base is 2. Thus a number can be written as $.f \times 2^e$. The number is stored in three parts: the sign, the exponent e, and the fraction f. In IEEE standard arithmetic, the sign uses 1 bit, the exponent uses 8 bits, and the fraction uses 23 bits. Without worrying about the details, the result is that real numbers may be stored with approximately 6 decimal digits of precision and the largest value is `huge(0.0)`, which is approximately 10^{38}.

Exercises

1. Use the intrinsic function `huge` to compute and print the largest value of each of the integer kinds available on your system.
2. Use the intrinsic functions `huge` and `precision` to compute and print the largest double precision value and the precision of double precision values on your system.
3. Investigate the intrinsic functions `range`, `radix`, and `digits` by using them in a program. Determine which apply to reals and which to integers. Appendix A may contain some information of interest.

7.2 NaN and Inf

One of the interesting things about the IEEE system of arithmetic is that it includes representations for "values" that are not ordinary numbers. For example, in ordinary arithmetic, the value of 7.7/0.0 is not defined. However, $7.7/x$ gets larger and larger as x gets closer and closer to 0.0. This is the basis for saying that the result of 7.7/0.0 is infinity, or `Inf`. Similarly, –7.7/0.0 is negative infinity or `-Inf`.

```
program inf
   implicit none
   print *, 7.7/0.0, -7.7/0.0
end program inf
```

```
 +Inf -Inf
```

There is no similar argument that will produce an answer for 0.0/0.0. $0.0/x$ is always 0.0, except when x is 0, so 0 might be a reasonable value. However, $x/0.0$ is infinity unless x is 0. On the other hand, x/x is always 1.0 unless x is 0.0. Thus 0.0/0.0 is considered to be "Not a Number" or `NaN`.

```
program nan
   implicit none
   print *, 0.0/0.0
end program nan
```

```
 NaN
```

As is the case for many programs in this book (the one later in this chapter that checks for overflow is an exception), it is assumed the program is run in "continue mode"—that is, the program does not halt after an exception occurs.

The reason any programmer might need to know about `Inf` and `NaN` is to understand why such a value gets produced. In most cases, this would indicate something wrong with the program and the value printed should provide some clue about what it is. For example, suppose a program is supposed to compute the roots of a quadratic equation $ax^2 + bx + c = 0$. If it does not seem to be working right and we try to debug the program by printing the value `sqrt(b**2-4*a*c)` and the result is `NaN`, that indicates that $b^2 - 4ac$ is probably negative and the roots are complex.

It is interesting, but predictable, how these two special values combine with ordinary arithmetic values. For example adding any ordinary number to `Inf` produces `Inf` and combining a `NaN` with anything produces another `NaN`.

```
program naninf

   implicit none
   real, parameter :: NaN = 0.0/0.0, &
                      Inf = 1.1/0.0

   print *, Inf + 4.4, Inf + Inf, Inf/5.5
   print *, 0*Inf, Inf - Inf, NaN + 4.4

end program naninf
```

```
 +Inf +Inf +Inf
 NaN NaN NaN
```

Other interesting results occur when two values are compared and one of them is `Inf` or `NaN`. The results of comparing `Inf` with ordinary values is not too surprising when you think of `+Inf` as bigger than any ordinary number and `-Inf` as smaller than any normal number.

```
program compare_inf
   implicit none
   real, parameter :: Inf = 4.4/0.0

   print *, 7.7 < Inf
   print *, 7.7 <= -Inf
   print *, Inf == Inf
   print *, -Inf < Inf

end program compare_inf
```

```
T
F
T
T
```

Comparing a NaN is more interesting. Basically the idea is that since NaN is not a number, it does not compare with any of the other values, so using both == and /= produce a false result. The only exception is that NaN /= NaN is true! This is shown in the example in the next subsection.

The *ieee_arithmetic* Intrinsic Module

This ieee_arithmetic intrinsic module has parameters and procedures to help manipulate IEEE arithmetic values. There is a procedure ieee_value that constructs the special values (otherwise, an exception would be raised) using parameter names from the module to indicate which value is desired. Note that there are two NaN values, one that signals an exception (7.3) when used and one that does not. It is better to use this procedure to construct Inf and NaN values.

```
program ieee_naninf

   use, intrinsic :: ieee_arithmetic
   implicit none
   real :: NaN, Inf

   NaN = ieee_value(0.0, ieee_quiet_nan)
   Inf = ieee_value(0.0, ieee_positive_inf)
   print *, Inf, NaN

end program ieee_naninf
```

```
 +Inf NaN
```

This program uses two of the several values from the ieee_arithmetic module. Please consult a reference manual or book to determine the others that are available.

Other procedures in the ieee_arithmetic module can be used to test values.

```
program ieee_test_naninf
   use, intrinsic :: ieee_arithmetic
   implicit none
   real :: NaN, Inf

   NaN = ieee_value(0.0,ieee_quiet_nan)
   Inf = ieee_value(0.0,ieee_positive_inf)

   print *, ieee_is_nan(NaN), ieee_is_nan(7.7), ieee_is_nan(Inf)
   print *, ieee_is_finite(7.7), ieee_is_finite(Inf)
   print *, Inf==NaN, NaN==NaN, NaN/=NaN
end program ieee_test_naninf
```

```
T F F
T F
F F T
```

Exercises

1. If `Inf` and `NaN` are given values as in the program `ieee_test_naninf`, what is the value of

   a. `NaN < Inf`

   b. `-Inf == -Inf`

2. If `Inf` and `NaN` are given values as in the program `ieee_test_naninf`, what is the value of `ieee_is_finite(0.0)`?

7.3 Exceptions

When a computation produces an unusual result, often it is acceptable to just have the program halt with a printed message that a problem has occurred. However, on other occasions, it might be important to have the program continue, check to see if there is a problem, and take corrective action. For examples, if the numbers get too large, all the data could be scaled down by a factor of 1000 and the calculations performed again. Usually these situations arise in programs written by professionals to perform sophisticated calculations such as solving a differential equation. Things can get complicated when using many of these features, but we will discuss a simple example in order to get an idea of the facilities available.

Suppose we suspect that a particular multiplication is producing an overflow—that is, it is producing a value bigger than the largest real value for the default real kind. One simple way to do this is to see if the product is `+Inf`, but we may not want to put this kind of checking into the program in order to check each time and possibly make some sort of computational adjustment if an overflow occurs. So what we do is take the essence of the computation and the checking and put it into a complete program in order to figure out how it all works.

The special procedures that are needed are accessible from the intrinsic module `ieee_exceptions`. The `use` statement in the program has an `only` clause, which lists just those features needed. This is nice not only for documentation of the program, but will produce a diagnostic if we misspell a name or try to get something from the wrong module. Also, using the procedure `ieee_get_flag` as an inquiry for the occurrence of the desired exception avoids the need to check each operation; this would be a laborious task in a large calculation.

```
program exception
   use, intrinsic :: ieee_exceptions, only: &
         ieee_overflow, ieee_get_flag, ieee_set_halting_mode
   implicit none

   real :: x, y, z
   logical :: overflow_flag = .false.

   print *, "Enter values for x and y to be multiplied:"
   read *, x, y

   ! Set the halting mode to continue even if there is overflow.
   call ieee_set_halting_mode(ieee_overflow, .false.)

   ! Do the multiplication
   z = x * y

   ! Check if overflow occurred
   call ieee_get_flag(ieee_overflow, overflow_flag)
   if (overflow_flag) then
      print *, "An overflow occurred in the product x*y"
   else
      print *, "No overflow occurred in the product x*y"
   end if
   print *, "    where x = ", x, " and y = ", y
   print *, "The product was computed as ", z
end program exception
```

After the type declarations, there are statements to print a prompt and read in values for the variables x and y. Before the two values are multiplied, the procedure ieee_set_halting_mode is called so that exception will be signaled if there is an overflow in a computation that follows. The value .false. indicates that the program is not to halt if there is an overflow.

After the multiplication is performed the module subroutine ieee_get_flag is called to determine if overflow occurred. The result is stored in the logical variable overflow_flag. Then the flag is tested and an appropriate message is printed. In the context of a larger program, action other than printing a message might be taken.

In a production program, more should be done. First, a check should be made that IEEE arithmetic is supported by the compiler. Then we should check that overflow checking is supported. We assume that both of these is true, but they can be checked using other features of the IEEE modules.

There are many other features of the IEEE modules, which can be discovered in a reference manual or book.

Exercise

1. Write a complete program that enables an exception for dividing by zero, does a divide by zero, then tests and reports the exception.

8 More about Modules and Procedures

This chapter covers submodules, elemental procedures, how to call C functions, and information about more intrinsic procedures.

8.1 Submodules

Modules (3.1) provide an excellent tool for organizing Fortran programs, but extensive use of modules has revealed some limitations that are solved by submodules. These difficulties arise particularly when large programs are constructed.

- Splitting a large module into several smaller modules causes private entities to be made public in order to be accessed by another one of the modules. This may also cause unnecessary name conflicts.
- Cascading compilation: when one module is changed, many others often are recompiled unnecessarily. This happens when the internals of a procedure change requiring its recompilation, but its interface does not change, so units dependent on that procedure do not need to be recompiled.
- Two modules may need to use entities from each other. (This is prohibited, but submodules can be used).

Interfaces in the module may be published without disclosing the implementation details because interfaces usually are placed in the module, but the implementation is in the submodule.

A submodule is a separately compiled program unit.

In this example, length is a type-bound procedure (6.2), but this is not essential; it could be an ordinary module procedure.

```
module line_mod

   implicit none
   private

   type, public :: line
      real :: x1, y1, x2, y2
```

```
   contains
      procedure :: length
   end type line

   interface
      module function length (l)
         class(line), intent(in) :: l
         real :: length
      end function length
   end interface

end module line_mod

submodule (line_mod) line_length_mod

contains

   module procedure length

      length = sqrt((l%x2-l%x1)**2 + &
                    (l%y2-l%y1)**2)

   end procedure length

end submodule line_length_mod

program submod

   use line_mod
   implicit none
   type (line) :: line_1

   line_1 = line(0, 0, 1, 1)
   print * line_1%length

end program submod
```

- Submodule procedures also are called **separate procedures**.
- Submodules access entities from the module by host association (just like with `contains`).
- Submodules may have local data.
- Submodules may have submodules.
- Submodules may access other modules via a `use` statement. A submodule of module A may use module B and a submodule of B may use A.

8.2 Date and Time Subroutines

There are three subroutines useful for getting information about the date and time: `date_and_time`, `system_clock`, and `cpu_time`.

Timing a Program

The following program illustrates a simple use of `date_and_time` to get the current date and `cpu_time` to compare the timings of two ways to do a matrix multiplication. One way uses `do` loops and the other uses the intrinsic function `matmul`. The timing is done by calling `cpu_time` before and after the code to be timed. The time spent executing the code is then the difference between the two values returned, given in seconds.

```
program time_matrix_multiply

   ! Compare times of the matmul intrinsic vs. DO loops

   implicit none
   integer, parameter :: n = 2000
   real, dimension(n, n) :: a, b, c1, c2
   character(len=8) :: date
   real :: start_time, stop_time
   integer :: i, j, k
   character(len=*), parameter :: form = "(t2, a, f0.3, a)"

   ! Get date to print on report

   call date_and_time(date=date)

   print *, "Timing report dated: " // &
      date(1:4) // "-" // date(5:6) // "-" // date(7:8)

   call random_seed()
   call random_number(a)
   call random_number(b)
   call cpu_time(start_time)
   c1 = 0
   do k = 1, n
      do j = 1, n
         do i = 1, n
            c1(i, j) = c1(i, j) + a(i, k) * b(k, j)
         end do
      end do
   end do
   call cpu_time(stop_time)
   print *
   print form, "Time of DO loop version is: ", &
      stop_time - start_time, " seconds."
```

```
   call cpu_time(start_time)
   c2 = matmul(a,b)
   call cpu_time(stop_time)

   print *
   print form, "Time of matmul version is: ", &
      stop_time - start_time, " seconds."

   print *
   if (any(abs(c1-c2) > 1.0e-4)) then
      print *, "There are significantly different values."
   else
      print *, "The results are approximately the same."
   end if

end program time_matrix_multiply
```

Here is one sample result of executing the program. Try this on your computer to see how your results differ.

```
Timing report dated: 2015-03-16

Time of DO loop version is: 6.297 seconds.

Time of matmul version is: 3.016 seconds.

The results are approximately the same.
```

Exercise

1. Write a program to time the generation of one million random numbers, first one at a time with one million calls to the intrinsic subroutine random_number, then with one call passing the whole array.

8.3 Command-Line Arguments

The intrinsic procedures command_argument_count, get_command, and get_command_argument are used to get information about the command that invoked the program. Suppose we have a debugging version of the program compute_it that prints out a lot of information about various variables during execution of the program, which the normal version does not. Which is to be run is indicated by putting one of the words debug or normal after the name of the program when executing it. The program tests the command and sets a logical variable print_debug_info accordingly.

```
program debug_it
   implicit none
   character(len=9) :: debug
   logical :: print_debug_info
      . . .
   call get_command_argument(number=1, value=debug)
   select case (trim(debug))
      case ("debug")
         print_debug_info = .true.
      case ("normal")
         print_debug_info = .false.
      case default
         print *, "Invalid command line argument"
         stop
   end select
      . . .
   if (print_debug_info) then
   ! print some debugging information
      . . .
end program debug_it
```

8.4 Environment Variables

The intrinsic subroutine `get_environment_variable` gets the value of the named environment variable.

```
character(len=99) :: path
   . . .
call get_environment_variable(name="PATH", value=path)
if (index(path, "my_dir") > 0) then . . .
```

8.5 Executing a System Command

The intrinsic subroutine `execute_command_line` may be used to execute any operating system command. This can be done synchronously or asynchronously and there are optional arguments to control this and other features. Here is a simple example. Suppose compiling the following program produces a file `stop_test.exe`. Note that the stop code now may be any integer or character constant expression.

```
program stop_test
   print *, "Preparing to stop"
   stop 999999
end program stop_test
```

Then a program can be written and run that executes stop_test or executes any other command.

```
program exec_command

   implicit none
   integer :: stop_code
   call execute_command_line("dir")
   call execute_command_line &
      ("stop_test.exe", exitstat=stop_code)
   print *, stop_code

end program exec_command
```

Running the program produces the result of executing the dir (Windows) system command (it would be ls on Linux) and the stop.exe program.

```
 Volume in drive C has no label.
 Volume Serial Number is BADA-0412

 Directory of C:\walt\training\F0308\Tests

11/19/2011  04:42 PM    <DIR>          .
11/19/2011  04:42 PM    <DIR>          ..
11/19/2011  04:08 PM            30,370 a.exe
11/19/2011  04:15 PM               212 command.f90
11/19/2011  04:40 PM            30,360 stop_test.exe
11/19/2011  04:40 PM                77 stop_test.f90
   . . .
 Preparing to stop
STOP 999999
      999999
```

8.6 Generic Procedures

Many intrinsic procedures are generic in that they allow arguments of different types. For example, the intrinsic function abs will take an integer, real, or complex argument. The programmer also can write generic procedures.

In 3.5 there is a subroutine that exchanges the values of any two real variables. It would be nice to have a similar routine that swapped integer values, but the normal rules of argument matching indicate that the types of the dummy and actual arguments must match. This is true, but it is possible to have one procedure name swap stand for several swapping routines, each with different names. The correct routine is picked for execution based on the types of the arguments, just as for generic intrinsic functions.

Here is the swap subroutine from 3.5, but with its name changed to swap_reals.

```
subroutine swap_reals(a, b)
   real, intent(in out) :: a, b
   real :: temp
   temp = a
   a = b
   b = temp
end subroutine swap_reals
```

It is easy to construct a similar subroutine swap_integers.

```
subroutine swap_integers(a, b)
   integer, intent(in out) :: a, b
   integer :: temp
   temp = a
   a = b
   b = temp
end subroutine swap_integers
```

The way to make them both callable by the generic name swap is to place the name swap in an interface statement and list the procedures that can be called when the arguments match appropriately. The result is an interface block that has a different form from the one used to declare a dummy procedure in 3.8.

```
public :: swap
private :: swap_reals, swap_integers
interface swap
   procedure swap_reals, swap_integers
end interface
```

When the interface block and the two subroutines are placed in a module, a program that uses the module can call swap with either two integer arguments or two real arguments. Here is the module and a program that tests the generic procedure swap.

```
module swap_module

   implicit none
   public :: swap
   private :: swap_reals, swap_integers

   interface swap
      procedure swap_reals, swap_integers
   end interface

contains

subroutine swap_reals(a, b)
   real, intent(in out) :: a, b
   real :: temp
```

```
      temp = a
      a = b
      b = temp
   end subroutine swap_reals

   subroutine swap_integers(a, b)
      integer, intent(in out) :: a, b
      integer :: temp
      temp = a
      a = b
      b = temp
   end subroutine swap_integers

   end module swap_module

   program test_swap

      use swap_module
      implicit none
      real :: x, y
      integer :: i, j

      x = 1.1
      y = 2.2

      i = 1
      j = 2

      call swap(x, y)
      print *, x, y

      call swap(i, j)
      print *, i, j

   end program test_swap
```

Running this program produces

```
      2.2000000    1.1000000
    2 1
```

Exercises

1. Extend the generic subroutine **swap** to handle arrays of integers.
2. Extend the generic subroutine **swap** to handle character strings.
3. Extend the generic subroutine **swap** to handle real values with precision greater than that of the default.

4. The program seven_11 in 3.15 calls the function random_int, which produces one pseudorandom integer value. The program seven_11 in 4.7 calls the function random_int, which produces an array of pseudorandom integer values. Write a module that makes random_int generic in the sense that it can be called to either produce a single integer value or an array of values, depending on its arguments.

8.7 Elemental Procedures

One way to extend the generic subroutine swap, as requested in Exercise 1 of 8.6, is to write another module procedure and add it to the list of procedures implementing the generic subroutine swap. A far easier solution is to make the subroutine swap_integers elemental.

An **elemental procedure** is one written with scalar (nonarray) dummy arguments, but which can be called with array actual arguments. When this is done, the computation in the procedure is performed element-by-element on each element of the array (or arrays) as if the invocation of the procedure were in a loop, executed once for each element of an array.

Here is how the generic subroutine swap can be made to apply to integer arrays and how it could be called to swap two arrays.

```
module swap_module

   implicit none
   public :: swap
   private :: swap_reals, swap_integers

   interface swap
      procedure swap_reals, swap_integers
   end interface

contains

elemental subroutine swap_reals(a, b)
   real, intent(in out) :: a, b
   real :: temp
   temp = a
   a = b
   b = temp
end subroutine swap_reals

elemental subroutine swap_integers(a, b)
   integer, intent(in out) :: a, b
   integer :: temp
```

```
      temp = a
      a = b
      b = temp
   end subroutine swap_integers

   end module swap_module

   program test_swap_arrays

      use swap_module
      implicit none
      integer, dimension(3) :: i = [1, 2, 3], &
                               j = [7, 8, 9]

      call swap(i, j)
      print *, i
      print *, j

   end program test_swap_arrays
```

```
 7 8 9
 1 2 3
```

Here are some rules for elemental procedures:

1. All of the dummy arguments must be scalar.
2. With a couple of unusual exceptions, the actual arguments must all be conformable.
3. An elemental procedure may not be recursive.
4. No dummy argument may be a pointer (10.1) and the result may not be a pointer.
5. No dummy argument may be a procedure.
6. Each elemental procedure must be pure unless the keyword `impure` is present.

Here is another example of an elemental function.

```
program test_elemental_function

   character(len=*), parameter :: format = "(3f7.2)"
   print format, f(1.1)
   print format, f([1.1, 2.2, 3.3])

contains
```

```
   elemental function f(x) result(rf)

      real, intent(in) :: x
      real :: rf

      rf = x**2 + 3

   end function f

end program test_elemental_function
```

```
   4.21
   4.21   7.84  13.89
```

Exercise

1. Rewrite the subroutine `random_int` in 4.7 so that it is elemental. It also must be declared impure because it calls `random_number`, which has a side effect.

```
impure elemental subroutine random_int(result, low, high)
```

8.8 More Array Intrinsic Procedures

There is a large set of intrinsic procedures that process arrays. Some of these have been discussed in Chapter 4. We will look at a few more and investigate some of the optional arguments allowed in some of them.

The simplest intrinsic functions that process arrays are elemental extensions of scalar versions of the functions, some of which have been in Fortran for a long time. Consider the trigonometric cosine function `cos`. It has long been generic in that it will take real and complex arguments, but it is also elemental in that the argument may be an array of any rank. For example, suppose we want to store in the variable `ss` the value of

$$\sum_{i=1}^{n} a_i \times \cos x_i$$

This can be done with the Fortran statement

```
ss = sum( a(1:n) * cos(x(1:n)) )
```

The cosine of each of the elements of `x` is computed and each of these, element-by-element, is multiplied by the appropriate element of `a`. The elements of the resulting array are added together by the `sum` intrinsic function. Note that `sum` is *not* elemental.

Suppose we want to add up the positive elements of the array and store the result as the value of the variable `sum_pos`. For this, `sum` has an optional argument (3.8) whose keyword is `mask`. The mask is a logical array conformable to the array `a` and only the elements of a for which the mask is true contribute to the sum.

```
sum_pos = sum (a, mask = (a > 0))
```

The operation `a > 0` elementally decides (true or false) whether a particular element is positive. This produces an array of logical values used as the mask.

Test Scores

Next, let us look at some functions that are specially designed to process arrays. Suppose we have a class with three students and have recorded their scores on four tests. These are stored in a 3 × 4 array named `score`, as follows:

$$\begin{bmatrix} 85 & 76 & 90 & 60 \\ 71 & 45 & 50 & 80 \\ 66 & 45 & 21 & 55 \end{bmatrix}$$

The largest score may be computed by `maxval(score)`, which is 90 in this case.

Suppose we want the three scores that are the largest score for each student. `maxval`, like many array intrinsic functions, has an optional argument with the keyword `dim`, which indicates that the operation should be performed by varying one of the subscripts of the array and keeping the others fixed. Thus, the expression

```
maxval(score, dim = 2)
```

computes the maximum of the four values in the first row (fix the first dimension to row 1 and vary the column number) as 90 and calculates the maximum value in the other rows to produce the array of three maximum values [90 80 66].

Now suppose we want to know which student got the largest score. We already have the array of the three largest scores [90 80 66]. The intrinsic function `maxloc` will determine the position in an array of the largest of the three scores, which is 1.

```
maxloc(maxval(score, dim = 2))
```

It may seem strange at first that the value of this expression is an array containing the single number 1. But if `maxloc` were applied to a two-dimensional array, what would be needed to locate the largest value is a row number and a column number—an array containing two numbers. The value of `maxloc(score)` is the array [1 3]. The first element of this array is the student with the largest score, giving us a second way of computing that. The second element is the test on which the largest score was achieved.

Average Score

Computing the overall average test score is easy. Divide the sum of all the scores by the number of scores, which is the size of the array.

```
average = sum(score) / size(score)
```

How many of the 12 scores are above the average, which is 62? First we produce an array of logical values, true where the score is above average and false where it is not. The variable above must be declared to be a 3 × 4 logical array.

```
above = (score > average)
```

Then the intrinsic function count (not elemental) determines the number of values in the array that are true. The answer happens to be 6 in our case.

```
qty_above_average = count(above)
```

We are going to use the logical array above elsewhere in the next example; otherwise, the two steps to compute qty_above_average could be combined.

```
qty_above_average = count(score > average)
```

Finally, just to show how powerful these intrinsics are, we ask the question: Did any student always score above the overall average? This is computed by the statement

```
any_student_always_above_average = any(all(above, dim = 2))
```

The intrinsic function all tests whether all of the elements of an array are true. With the dim optional argument 2, it tests each row for all true values. This produces three false values. Then the function any tests if any of the values in its array argument is true. The answer is false in our case. The first student came closest, scoring above the average 62 on all but the last test.

The findloc function

The findloc function can be used to search an array for a specific value. Because it is never a good idea to test real values for an exact match, it seems that many uses of the function will temporarily create an array of logical values, then search that array for a value that is true.

```
program find_loc

   implicit none
   intrinsic :: findloc

   real, dimension (3,3) :: x = &
      reshape (                 &
```

```
        [ -11,  12, -13,        &
           21,  22, -23,        &
           31, -32, -33 ],      &
        [ 3, 3 ], order = [ 2, 1 ] )

   logical, parameter :: T = .true.

   print *, findloc(X>0, T)          ! = [ 2, 1 ]
   print *, findloc(X>0, T, back=T)  ! = [ 2, 2 ]
   print *, findloc(X>0, T, dim=2)   ! = [ 2, 1, 1 ]

end program find_loc
```

Exercise

1. Write a statement that computes the answer to the question: Was any test so easy that all four students scored above average on that test?

8.9 Bit Intrinsic Procedures

Sometimes it is convenient to be able to manipulate the individual binary digits (bits) of a Fortran integer value. One example might involve using the bits of a large array of integers to represent the states of the components of an electronic circuit. The following example uses only two integer values to store and manipulate bits just to see how things work.

The bits of integer value are numbered right to left starting with bit 0 on the right; the integer is assumed to be stored using a binary representation. The first executable statement of the program starts with integer value 0, whose bits are all 0, and sets bit 3 to 1 using the intrinsic function `bset`. Then bit two is also set by immediately calling `bset` again and the result is saved as the value of `b1100`. Since the decimal equivalent of binary 1100, which has bits 3 and 2 set, is 12, the result of this assignment is the same as setting the variable `b1100` to 12. Similarly, bits 3 and 1 of `b1010` are set, resulting in `b1010` having the value 10.

This is all verified by printing `b1100` as an integer and using the intrinsic function `btest` to check which of the bits 3, 2, 1, and 0 of `b1100` and `b1010` are set. Similar `print` statements are used to show the values of `ior` and `iand` applied to these two values.

```
program bits

   implicit none
   integer :: b1100, b1010
   character(len=*), parameter :: &
      form = "(a15, 4l2)"
   integer :: k
```

```
   b1100 = ibset(ibset(0,3),2)
   b1010 = ibset(ibset(0,3),1)

   print *, "The integer value of b1100 is", b1100
   print *

   print form, "b1100", &
      (btest(b1100, k), k = 3, 0, -1)
   print form, "b1010", &
      (btest(b1010, k), k = 3, 0, -1)
   print form, "Logical or", &
      (btest(ior(b1100,b1010), k), k = 3, 0, -1)
   print form, "Logical and", &
      (btest(iand(b1100,b1010), k), k = 3, 0, -1)

end program bits
```

Each list of data to be printed looks a lot like a do loop. It is called an **implied-do loop**. It works like a do loop, printing the value indicated for k equal to 3, 2, 1, and 0. Note that the whole loop is surrounded by parentheses.

Running this program produces output that illustrates how *logical or* and *logical and* are computed.

```
The integer value of b1100 is 12

          b1100 T T F F
          b1010 T F T F
     Logical or T T T F
    Logical and T F F F
```

The other bit intrinsic procedures are ibclr, which sets a bit to 0, ieor, which computes exclusive or, ishft and ishftc which perform end-off and circular shifts, not, which complements the bits of its argument, and the mvbits subroutine, which copies bits from one integer to another.

Exercise

1. Write a function number_of_bits(n) that counts the number of bits in an integer argument n that are 1. Test the function by printing the resulting values for n = –16, –15, ..., –2, –1, 0, 1, 2, ..., 16. Also print the values of the function when called with arguments huge(n) and –huge(n).

8.10 Calling C Procedures

Sometimes, it is necessary or desirable for a Fortran program to call a procedure written in another programming language. For example, the procedure may be written already and save you some programming effort; it may use features of another language that are more convenient than or are missing from Fortran, such as interaction with a game screen or a telemetry data stream.

There are several problems that make it a little difficult to correctly call a C procedure. The most obvious is that the actual Fortran and dummy C arguments must match. We cannot say that the types must be the same because data types have different names in C (`int` instead of `integer`, for example). What must be true is the *representations* must be the same. The different representations of data of a specific type in Fortran are indicated by kinds; therefore, the problem is to use corresponding types and select the right kind for the Fortran data. Another problem is simply getting the name of the C procedure right as the system may use something different than the name given by the C programmer; a typical case is that systems put an underscore after procedure names when building the whole program, and so the C procedure `print_a_line` must be called `print_a_line_`. However, these conventions are different on different computer systems. There are special features in Fortran to make it a little easier to call a C procedure on any system using exactly the same Fortran program and requiring no knowledge of these C and system idiosyncrasies.

Interfaces to C Procedures

To have a Fortran program make a correct call to a C procedure, an interface describing the C procedure is written in the calling Fortran program. Procedure interfaces are described in 3.8. The main new feature that is needed is to put `bind(c)` at the end of the subroutine or function statement in the interface describing a C procedure. This tells the Fortran compiler that it is a C procedure being called, even though the interface information is all written in Fortran.

C has only functions, no subroutines. However, a function can be declared as returning `void`, a special indication that the function does not return a regular value; such a function should be called as a subroutine from Fortran, whereas all other C functions should be called as Fortran functions.

The *iso_c_binding* Intrinsic Module

There is a built-in (intrinsic) module named `iso_c_binding`. All you have to do is use it to have access to several named constants and procedures that will help create a correct call to a C procedure, on whatever system the program is run. The keyword `intrinsic` can be added to the `use` statement, as shown in the example below, to ensure that you access the intrinsic module, not a user-written module that happens to have the same name.

One of the main features of this intrinsic module is a collection of named constants (parameters) that are Fortran kind numbers that indicate kinds that correspond to C data types. C does not use kinds, but uses a different data type for each kind, so that, for example, the C data types `float` and `double` correspond to two different kinds of `real` in Fortran. Table 8-1 lists a few of these parameters for the most common C data types.

Table 8-1 C types and Fortran kinds

C type	Fortran type and type parameter
`int`	`integer(kind=c_int)`
`short int`	`integer(kind=c_short)`
`long int`	`integer(kind=c_long)`
`float`	`real(kind=c_float)`
`double`	`real(kind=c_double)`
`char`	`character(len=1, kind=c_char)`

C does not have the character string as a data type. Strings are represented by arrays of single characters. However, a Fortran program may call a C function with a dummy argument that is an array of characters and pass a character string actual argument. This is one of the things that is taken care of by putting `bind(c)` on the interface of the procedure being called. Note that in the interface, the dummy character array is declared to have `dimension(*)` (do not ask why—just do it).

Another parameter in the `iso_c_binding` intrinsic is `c_null_char`. It is a single character that is used to terminate all C strings and hence needs to occur at the end of a character string passed to a C function.

The `iso_c_binding` intrinsic module contains other things useful for interoperating with C.

An Example of Interoperation with C

The following example illustrates the use of some of these features. The example shows how to pass a character string, an array, and a structure, as well as simple variables.

```
module type_def

   use, intrinsic :: iso_c_binding
   implicit none
   private
```

```
   type, public, bind(c) :: t_type
      integer(kind=c_int) :: count
      real(kind=c_float) :: data
   end type t_type

end module type_def

program fortran_calls_c

   use type_def
   use, intrinsic :: iso_c_binding
   implicit none

   type(t_type) :: t
   real(kind=c_float) :: x, y
   integer(kind=c_int), dimension(0:1, 0:2) :: a

   interface
      subroutine c(tp, arr, a, b, m) bind(c)
         import :: c_float, c_int, c_char, t_type
         type(t_type) :: tp
         integer(kind=c_int), dimension(0:1, 0:2) :: arr
         real(kind=c_float) :: a, b
         character(kind=c_char), dimension(*) :: m
      end subroutine c
   end interface

   t = t_type(count=99, data=9.9)
   x = 1.1
   a = reshape([1, 2, 3, 4, 5, 6], shape(a))
   call c(t, a, x, y, "doubling x" // c_null_char)
   print *, x, y
   print *, t
   print *, a

end program fortran_calls_c
```

Following is the C program that implements the function c.

```
typedef struct {int amount; float value;} newtype;

void c(newtype *nt, int arr[3][2], float *a, float *b, char msg[])
{
   printf (" %d %f\n", nt->amount, nt->value);
   printf (" %d %d %d\n", arr[0][1], arr[1][0], arr[1][1]);
   printf (" %s\n", msg);
   *b = 2*(*a);
}
```

Following is the output:

```
99 9.90000
2 3 4
doubling x
1.1 2.2
99 9.9
1 2 3 4 5 6
```

The module defines the derived type `t_type`. A structure `tp` of this type is passed as an actual argument, so the components must be declared to agree with the structure `nt` of type `newtype` in the C function. This uses the parameters `c_int` and `c_float` from the intrinsic module. The type itself is defined with the `bind(c)` attribute so that the components will be laid out in a manner similar to the components of a C structure of type `newtype`.

The program `fortran_calls_c` uses the module `type_def` to access the derived type `t_type`. The structure `t` is declared to be `t_type`. `x` and `y` are real variables with the kind to match a C `float` data type.

All C arrays have lower bound 0 and are stored by rows rather than by columns, as Fortran arrays are stored. Hence, it makes it a little easier to give the array lower bounds 0. Also note that the subscripts are reversed in the declaration of the Fortran actual array `a` and the C dummy array `arr`.

An interface for the C function `c` includes `bind(c)` and it uses the `import` statement to make some of the parameters in the intrinsic modules accessible within the interface, which has its own scope (3.13).

`t` is given a value with a structure constructor (6.3), `x` is given the value 1.1, and `a` is given the value

$$\begin{bmatrix} 1 & 3 & 5 \\ 2 & 4 & 6 \end{bmatrix}$$

with an array constructor (4.1) and the `reshape` function (4.1).

The C procedure is called as a subroutine and values are printed in both the C procedure and the main program.

There are many more features involved with the interoperability with C; for example, pointers can be passed and a C function can call a Fortran procedure under the right circumstances. If you need to interact with C in a more complicated manner, please consult a manual or reference work, such as *The Fortran 2003 Handbook*.

9 Extending Fortran

In Fortran, the programmer may not only define new types (6.2), but may define new operators and extend the definition of intrinsic functions, existing operators, and assignment. This allows the user to define a special environment for each application. In this chapter, we first see how to write code to do all of these things, then apply the techniques to a useful new data type consisting of big integer values.

9.1 Extending Assignment

When an assignment statement is executed, sometimes the data type of the expression on the right-hand side of the assignment symbol (=) is converted to the type of the variable on the left-hand side. For example, if `i` is integer and `r` is real, the assignment

```
r = i
```

causes the integer value of `i` to be converted to type real for assignment to `r`. Suppose we would like to extend this feature so that a logical value can be assigned to an integer with a false value being converted to zero and a true value converted to one when the assignment

```
i = log
```

is written with `log` logical type and `i` integer type. To do this, a subroutine that does the assignment must be written and an interface block must be given that indicates which subroutine does the assignment with conversion. Both these things should be placed in a module. The subroutine that will do the conversion follows.

```
subroutine integer_gets_logical(i, logical_expression)

   integer, intent(out) :: i
   logical, intent(in) :: logical_expression
```

```
    if (logical_expression) then
       i = 1
    else
       i = 0
    end if

 end subroutine integer_gets_logical
```

The following interface block indicates that assignment is extended by the subroutine `integer_gets_logical`. The `public` statement indicates that the extended assignment is available when the module is used, and the `private` statement indicates that the procedure `integer_gets_logical` is not accessible outside the module.

```
    public :: assignment(=)
    private :: integer_gets_logical

    interface assignment(=)
       procedure integer_gets_logical
    end interface
```

Here is the complete module to accomplish this task with a program that tests it.

```
 module int_logical_module

    implicit none
    public :: assignment(=)
    private :: integer_gets_logical

    interface assignment(=)
       procedure integer_gets_logical
    end interface

 contains

 subroutine integer_gets_logical(i, l)

    integer, intent(out) :: i
    logical, intent(in) :: l

    if (l) then
       i = 1
    else
       i = 0
    end if

 end subroutine integer_gets_logical

 end module int_logical_module
```

```
program test_int_logical

   use int_logical_module
   implicit none
   integer :: i, j

   i = .false.
   print *, i
   j = (5 < 7) .and. (sin(0.3) < 1.0)
   print *, j

end program test_int_logical
```

```
0
1
```

A subroutine that serves to define an assignment must have exactly two arguments; the first must be intent `out` or `in out` and the second intent `in`.

This mechanism must not be used to change the meaning of assignment with intrinsic types for which assignment is already defined, except that it may be used to redefine assignment between values of the same derived type. It also may be used to define assignment between values of different derived types.

Exercise

1. Write a module with a procedure and an interface block that extends assignment to allow assigning an integer to a logical variable. The logical variable should be set to false if the integer is 0 and set to true otherwise. Test the procedure with a program that uses the module.

9.2 Extending Operators

Suppose we now want to be able to use + in place of `.or.`, * in place of `.and.`, and - in place of `.not.` to manipulate logical values. This can be done by extending these operators, which already work with numeric operands. Functions must be written and the names of the functions placed in an interface block in a module. The interface statement contains the keyword `operator` in this case. Here is a complete module and program to implement and test the extension of + to logical operands. It would be reasonable to put the extensions for * and - in the same module.

```
module logical_plus_module

   implicit none
   public :: operator(+)
   private :: log_plus_log
```

```
   interface operator(+)
      procedure log_plus_log
   end interface

contains

function log_plus_log(x, y)  &
      result(log_plus_log_result)

   logical, intent(in) :: x, y
   logical :: log_plus_log_result

   log_plus_log_result = x .or. y

end function log_plus_log

end module logical_plus_module

program test_logical_plus

   use logical_plus_module
   implicit none

   print *, .false. + .false.
   print *, .true. + .true.
   print *, (2.2 > 5.5) + (3.3 > 1.1)
   print *, (2.2 > 5.5) .or. (3.3 > 1.1)

end program test_logical_plus
```

```
 F
 T
 T
 T
```

Note that the parentheses in the expression in the third `print` statement are necessary because + has a higher precedence than >. They are not necessary in the fourth `print` statement because `.or.` has lower precedence than >.

A function used to extend an operator must have one or two arguments (depending on the operator being extended), which must be intent `in`.

It is not allowed to use this mechanism to change the meaning of an existing operator applied to one of the intrinsic types for which it is defined; however, a new operator may be used for this purpose (9.3).

Exercises

1. Extend the module above to allow the operations of * and - with logical operands.

2. Write a module that extends the == and /= operators to allow comparison of both scalar logical values and arrays of logical values. Note that the built-in operators `.eqv.` and `.neqv.` are used for this purpose.

9.3 User-Defined Operators

In addition to extending the meaning of the built-in operators, it is possible to make up new names for operators. If we were to add the operation of testing if an integer is prime, there is probably not a good unary built-in operator that would be suitable to extend to this use. Any name consisting of from 1 to 63 letters preceded and followed by a period may be used, except that the operator name cannot be either of the logical constants `.true.` or `.false.` For example, we might pick `.prime.` for the name of the operator that returns true or false depending on whether its operand is a prime integer. Defining a new operator is similar to extending an existing one; its name is used in an interface statement and the function, which must have one or two intent `in` arguments, is named in a `procedure` statement.

```
interface operator (.prime.)
   procedure prime
end interface operator (.prime.)
```

This operator could now be used just like any built-in unary operator, as illustrated by the following `if` statement:

```
if (.prime. b .and. b > 100) print *, b
```

The precedence of a defined binary operator is always lower than all other operators, and the precedence of a defined unary operator is always higher than all other operators. Therefore, in the example above, `.prime.` is evaluated before `.and.`

Exercise

1. Implement and test the operator `.prime.` described in this section.

9.4 Extending Intrinsic Functions

Many programmers are surprised that the `sqrt` function may be used with a real or complex argument, but not with an integer argument. One possible reason is that there might be some controversy about whether the result should be an integer or real value. For example, should `sqrt(5)` be 2.236068, a type real approximation to the square root, or 2, the largest integer less than the real square root? The integer square root is sometimes useful; one example is in determining the upper bound on factors of an integer `i`.

It is not hard to compute either value with the expressions `sqrt(real(i))` and `int(sqrt(real(i)))` for any integer i, but it would be nice to just write `sqrt(i)`. We will extend the `sqrt` function to take an integer argument and return an integer value. This is done by writing an interface block and the function to do the computation. Here is the interface, the function, and a brief testing program. In the program, the 0.5 is used to avoid problems with roundoff.

Putting the keyword `elemental` on the function statement creates an array version for each of the ranks. It is called with a one-dimensional array in the program `test_integer_sqrt`.

Unlike with operators, it is possible to change the meaning of an intrinsic function definition for one type of argument without affecting the availability of that intrinsic for other types. For example, it is possible to change the definition of the cosine function `cos` for double precision arguments, but not affect the intrinsic definition for complex.

```
module integer_sqrt_module

   implicit none
   intrinsic :: sqrt
   public :: sqrt
   private :: sqrt_int

   interface sqrt
      procedure sqrt_int
   end interface

contains

elemental function sqrt_int(i) result(sqrt_int_result)

   integer, intent(in) :: i
   integer :: sqrt_int_result

   sqrt_int_result = int(sqrt(real(i) + 0.5))

end function sqrt_int

end module integer_sqrt_module

program test_integer_sqrt

   use integer_sqrt_module
   implicit none

   integer :: k
   integer, dimension(20) :: n = [ (k, k = 1, 20) ]
```

```
      print "(20i3)", n, sqrt(n)

   end program test_integer_sqrt
```

```
  1  2  3  4  5  6  7  8  9 10 11 12 13 14 15 16 17 18 19 20
  1  1  1  2  2  2  2  2  3  3  3  3  3  3  3  4  4  4  4  4
```

An intrinsic function should be listed in an `intrinsic` statement in a module containing a description of an extension.

A better way to compute the square root of an integer is with an iterative technique called "Newton–Raphson", which is used in the big integer module later in this chapter.

Exercises

1. Extend the intrinsic subroutine `random_number` so that it has the functionality of `random_int` in the previous exercise when called with arguments that are type integer. Allow both scalar and array arguments.

2. Modify the extended intrinsic subroutine `random_number` of the previous exercise so that the arguments `low` and `high` are optional as was done for `random_int` in Exercise 5 of 3.15.

9.5 Derived-Type Editing

A value of derived type may be read or written using edit descriptors for each of its components. This is not always the best format for a derived-type value. It is possible to write a subroutine that performs the input/output in whatever form is desired. However, it is convenient to simply put the derived-type value in an input/output list and have a single `dt` edit descriptor to specify the format. This is done in a manner analogous to extending assignment or extending an operator.

The form of the `dt` edit descriptor is

`dt` [*character-literal-constant*] [(*arg-list*)]

Using list-directed (* format) input/output with a derived-type object when a `dt` edit descriptor is available also causes the editing extension to be used to format the value. This is what is done in the examples below.

To start with a simple example, suppose the derived type `point` represents a point in the plane given by rectangular coordinates x and y.

```
type, public :: point
   real :: x, y
end type point
```

Suppose it is desirable to have values of type point printed as shown in the example:

```
print *, "The distance from ", p1, " to ", p2, " is ", distance(p1, p2)
```

where the printed result is

```
The distance from (5.0, 7.0) to (6.0, 8.0) is 1.414
```

Similar to extending an operator or assignment, an interface is used, which references a subroutine that does the output. Things can get pretty complicated when using all the features of derived-type editing, but our example will be about as simple as possible. First we need an interface.

```
interface write (formatted)
   procedure print_point
end interface
```

In place of write (formatted), the interface statement could contain read (formatted), write (unformatted), or read (unformatted), depending on the type of data transfer.

The subroutine print_point has a number of arguments that are required in order to handle more complicated examples; they are simply ignored in our example, but must be in the subroutine. The actual argument passed to iotype is the character string "DT"//*type*, where *type* is the character literal constant of the dt edit descriptor; iotype is LISTDIRECTED if the format is (*). The actual argument passed to v_list is the integer array of values making up the argument list that is the second optional part of the dt edit descriptor. unit, iostat, and iomsg are used in the same manner as the namesake keywords in other data transfer statements.

The arguments iotype and v_list are not used for unformatted input/output.

The interface and the subroutine will be placed in the module point_module and, of course, to make the example above work, a distance function is needed. Here is the module as constructed so far; it is reasonable to enhance the module with other operations to manipulate points.

```
module point_module

   implicit none
   private
   type, public :: point
      real :: x, y
   end type point

   interface write (formatted)
      procedure print_point
   end interface
   public :: write (formatted)
   public :: distance
```

```
contains

subroutine print_point(p, unit, iotype, v_list, iostat, iomsg)

   ! Arguments required by the dt edit descriptor
   type(point), intent(in) :: p
   integer, intent(in) :: unit
   character(len=*), intent(in) :: iotype
   integer, intent(in), dimension(:) :: v_list
   integer, intent(out) :: iostat
   character(len=*), intent(in out) :: iomsg

   write "(a, f0.1, a, f0.1, a)", "(", p%x, ", ", p%y, ")"
   iostat = 0

end subroutine print_point

function distance(p1, p2) result(d_r)
   type (point), intent(in) :: p1, p2
   real :: d_r
   d_r = sqrt ((p1%x-p2%x)**2 + (p1%y-p2%y)**2)
end function distance

end module point_module

program test_point_dt
   use point_module
   implicit none
   type (point) :: p1, p2
   p1 = point(5.0, 7.0)
   p2 = point(6.0, 8.0)
   print *, "The distance from ", p1, " to ", p2, " is ", &
              distance(p1, p2)
end program test_point_dt
```

9.6 Case Study: Computing with Big Integers

Suppose we are interested in adding, multiplying, and dividing very large integers, possibly with hundreds of digits. This kind of capability is needed to factor large integers, a task very important in cryptography and secure communications. The Fortran intrinsic integer type has a limit on the size of numbers it can represent, which depends on the system, but might be $2^{63} - 1$, which is 9,223,372,036,854,775,807. This problem can be solved by creating a new data type, called `big_integer`, deciding which operations are needed, and writing procedures that will perform the operations on values of this type. All of this will be placed in a module called `big_integers` so that it can be used by many programs.

The Type Definition for Big Integers

The first task is to decide how these large integers will be represented. Although a linked list of digits is a possibility, it seems more straightforward to use an array of ordinary Fortran integers. The only remaining thing to decide is how much of a big integer to put into each element of the array. One possibility would be to put as large a number into each element as possible. To make it easier to conceptualize with simple examples, we will store one decimal digit in each element. *However, because the abstract data type paradigm is followed, changing the representation so that larger integers are stored in each array element can be implemented easily without changing the programs using the* `big_integer` *module.*

The following type definition does the job. It uses a parameter `nr_of_digits` that has arbitrarily been set to 100; this allows decimal numbers with up to 100 digits to be represented using this scheme. The parameter `nr_of_digits` has the `private` attribute, which means it cannot be accessed outside the module.

```
integer, parameter, private :: nr_of_digits = 100

type, public :: big_integer
   private
   integer, dimension(0:nr_of_digits) :: digit
end type big_integer
```

The array `digit` has 101 elements. `digit(0)` holds the units digit; `digit(1)` holds the tens digit; `digit(2)` holds the hundreds digit; and so on. The extra element in the array is used to check for overflow—if any value other than zero gets put into the largest element, that will be considered to exceed the largest `big_integer` value and will be set to infinity.

The `private` statement indicates that we do not want anybody who uses the module to be able to access the *component* `digit` of a variable of type `big_integer`; we will provide all of the operations necessary to compute with such values. The `private` statement is discussed in 3.1.

IEEE arithmetic (7.2) deals only with real values. But just to get a little better grasp on some of the simple concepts, we will introduce the special values of Infinity and NaN into our scheme of big integers. These two special values will be represented by a 1 and 2 in position 100, with zeros elsewhere. These two values can be set up as parameters.

```
integer, private :: k
integer, private, parameter :: bInf = 1, bNaN = 2
type (big_integer), public, parameter :: &
      big_Inf = big_integer( [(0, k=0,nr_of_digits-1), bInf] ), &
      big_NaN = big_integer( [(0, k=0,nr_of_digits-1), bNaN] )
```

Here is an excellent example of the use of the `private` attribute. There is no reason for a user of the big integer module to have access to `k`, `bInf`, or `bNaN`.

Printing Big Integers

The next thing to do is to define some operations for big integers. The first necessary operations assign values to a big integer and print the value of a big integer. Let us take care of the printing first.

The following subroutine prints the value of a big integer. It takes advantage of the fact that each element of the array `digit` is a single decimal digit. This subroutine `print_big` is inside the module `big_integers` and so has access to all the data and procedures in the module.

```
subroutine print_big(b)

   type(big_integer), intent(in) :: b
   integer :: n, first_significant_digit

   if (b%digit(nr_of_digits) == bInf) then
      print "(a)", "Inf"
      return
   else if (b%digit(nr_of_digits) == bNaN) then
      print "(a)", "NaN"
      return
   end if

   ! Find first significant digit
   first_significant_digit = 0  ! In case b = 0
   do n = nr_of_digits, 1, -1
      if (b%digit(n) /= 0) then
         first_significant_digit = n
         exit
      end if
   end do

   print "(50i1)", b%digit(first_significant_digit:0:-1)

end subroutine print_big
```

The basic strategy is to print the digits in `i1` format, but first the leftmost nonzero digit must be located to avoid printing long strings of leading zeros.

Assigning Big Integers

To test the printing subroutine, we need to have a way to assign a value to a big integer. One possibility is to write a procedure that will assign an ordinary Fortran integer to a big integer, but this will limit the size of the integer that can be assigned. A second possibility is to write the integer as a character string consisting of only digits 0–9 (we are not allowing negative numbers). This is done by the subroutine `big_gets_char(b, c)` that assigns the integer represented by the character string `c` to the big integer b. If c contains a character other than one of the digits, the subroutine returns big_NaN. big_Inf is returned for b if the character string c is too long.

```
subroutine big_gets_char(b, c)

   type(big_integer), intent(out) :: b
   character(len=*), intent(in) :: c
   integer :: n, i
   character(len=*), parameter :: decimal_digits = "0123456789"

   if (verify(c, decimal_digits) /= 0) then
      b = big_NaN
      return
   else if (len(c) > nr_of_digits) then
      b = big_Inf
      return
   end if

   b%digit = 0
   n = 0
   do i = len(c), 1, -1
      b%digit(n) = index(decimal_digits, c(i:i)) - 1
      n = n + 1
   end do

end subroutine big_gets_char
```

Putting the Procedures in a Module

Now that we have enough operations defined on big integers to at least try something meaningful, we next need to package them all in a module. The module that we have created so far follows:

```
module big_integers

   implicit none

   integer, parameter, private :: nr_of_digits = 100

   type, public :: big_integer
      private
      integer, dimension(0:nr_of_digits) :: digit
   end type big_integer

   integer, private :: k
   integer, private, parameter :: bInf = 1, bNaN = 2
   type (big_integer), parameter :: &
         big_Inf = big_integer( [(0, k=0,nr_of_digits-1), bInf] ), &
         big_NaN = big_integer( [(0, k=0,nr_of_digits-1), bNaN] )
```

```
   public :: print_big, big_gets_char

contains

subroutine print_big(b)

   type(big_integer), intent(in) :: b
   integer :: n, first_significant_digit

   if (b%digit(nr_of_digits) == bInf) then
      print "(a)", "Inf"
      return
   else if (b%digit(nr_of_digits) == bNaN) then
      print "(a)", "NaN"
      return
   end if

   ! Find first significant digit
   first_significant_digit = 0  ! In case b = 0
   do n = nr_of_digits, 1, -1
      if (b%digit(n) /= 0) then
         first_significant_digit = n
         exit
      end if
   end do

   print "(50i1)", b%digit(first_significant_digit:0:-1)

end subroutine print_big

subroutine big_gets_char(b, c)

   type(big_integer), intent(out) :: b
   character(len=*), intent(in) :: c
   integer :: n, i
   character(len=*), parameter :: decimal_digits = "0123456789"

   if (verify(c, decimal_digits) /= 0) then
      b = big_NaN
      return
   else if (len(c) > nr_of_digits) then
      b = big_Inf
      return
   end if

   b%digit = 0
   n = 0
```

```
      do i = len(c), 1, -1
         b%digit(n) = index(decimal_digits, c(i:i)) - 1
         n = n + 1
      end do

   end subroutine big_gets_char

   end module big_integers
```

With the module available, we can write a simple program to test the assignment and printing routines for big integers.

```
   program test_big_1

      use big_integers
      implicit none
      type(big_integer) :: b1

      call big_gets_char(b1, "7123456789098765432l")
      call print_big(b1)
      print *

      call big_gets_char(b1, "")
      call print_big(b1)
      print *

      call big_gets_char(b1, "123456789+987654321")
      call print_big(b1)
      print *

   end program test_big_1
```

```
   71234567890987654321
   0
   NaN
```

Extending Assignment for Big Integers

The name for the subroutine `big_gets_char` was picked because it converts a character string to a big integer. But this is just like intrinsic assignment that converts an integer to a real value when necessary. Indeed, it is possible to use the assignment statement to do the conversion from character to type big integer. It is done by extending assignment as described in 9.1.

```
   interface assignment(=)
      procedure big_gets_char
   end interface
   public :: assignment(=)
   private :: big_gets_char
```

Now any user of the module can use the assignment statement instead of calling a subroutine, which makes the program a lot easier to understand.

Derived-Type Input/Output for Integers

The subroutine `print_big` can be fixed so that it extends formatted input/output to values of type `big_integer`, as was done for values of type `point` in 9.5. The only real work is to put in all of the required dummy arguments, but the executable code is the same as above. The declaration part of `print_big` is:

```
subroutine print_big(b, unit, iotype, v_list, iostat, iomsg)

   ! Arguments required by the dt edit descriptor
   type(big_integer, intent(in) :: b
   integer :: intent(in) :: unit
   character(len=*), intent(in) :: iotype
   integer, intent(in), dimension(:) :: v_list
   integer, intent(in out) :: iostat
   character(len=*), intent(in out) :: iomsg

   ! Local variables
   integer :: first_dignificant_digit, n
      . . .
```

A `write(formatted)` interface also must be created. These two extensions may be tested with the program `test_big_2`.

```
program test_big_2

   use big_integers
   implicit none
   type(big_integer) :: b1

   b1 = "71234567890987654321"
   print *, b1

   b1 = ""
   print *, b1

   b1 = "123456789+987654321"
   print *, b1

end program test_big_2
```

The result of running this version is identical to the output of `test_big_1`.

There is no need to have the subroutines `big_gets_char` and `print_big` available. They may be declared `private`.

Adding Big Integers

Now that we can assign to a big integer variable and print its value, it would be nice to be able to perform some computations with big integers. Addition can be done with a function that adds just like we do with pencil and paper, adding two digits at a time and keeping track of any carry, starting with the rightmost digits. The function `big_plus_big` does this.

```
pure function big_plus_big(x, y) result(big_plus_big_result)

   type(big_integer), intent(in) :: x, y
   type(big_integer) :: big_plus_big_result
   integer :: carry, temp_sum, n

   if (x==big_NaN .or. y==big_NaN) then
      big_plus_big_result = big_NaN
      return
   else if (x==big_Inf .or. y==big_Inf) then
      big_plus_big_result = big_Inf
      return
   end if

   carry = 0
   do n = 0, nr_of_digits
      temp_sum = x%digit(n) + y%digit(n) + carry
      big_plus_big_result%digit(n) = modulo(temp_sum, 10)
      carry = temp_sum / 10
   end do

   if (big_plus_big_result%digit(nr_of_digits) /= 0 &
         .or. carry /= 0) then
      big_plus_big_result = big_Inf
   end if

end function big_plus_big
```

In mathematics, the symbols + and − are used to add and subtract integers. It is nice to do the same with big integers, and it is possible to do so by extending the generic properties of the operations already built into Fortran. Note that + already can be used to add two integers, two real values, or one of each. The intrinsic operator + also can be used to add two arrays of the same shape. In that sense, addition is already generic. We now extend the meaning of this operation to our own newly defined type, `big_integer`. This is done with another interface block, this time with the keyword `operator`, followed by the operator being extended. The + operator is public, but the function `big_plus_big` is private. The function is explicitly `pure` because it will be called from a pure subroutine.

```
public :: operator (+)
interface operator (+)
   procedure big_plus_big
end interface
private :: big_plus_big
```

The use of the plus operator to add two big integers is tested by the program `test_big_3`.

```
program test_big_3

   use big_integers_module
   type(big_integer) :: b1, b2

   b1 = "12345678909876543210"
   b2 = "98765432101234567890"
   print *, b1 + b2

end program test_big_3
```

The output is

```
111111111011111111110
```

The operation `b1==b2`, where `b1` and `b2` are big integers, is not built in. It must be defined by a logical-valued function and placed in the module with the appropriate interface. This operation is used in the function `big_plus_big` to test if either of the operands is `Inf` or `NaN`. This is a little more robust than checking if digit 100 is 1 or 2. Writing the function to extend == is an exercise.

Using only the procedures written so far, it is not possible to use the expression `b + i` in a program where `b` is a big integer and `i` is an ordinary integer. To do that, we must write another function and add its name to the list of functions in the interface block for the plus operator. Similarly, it would be necessary to write a third function to handle the case `i + b`, because the arguments are in the reverse order of the function that implements `b + i`. Even if that is not done, the number 999 could be added to `b` using the statements

```
temp_big_integer = "999"
b = b + temp_big_integer
```

Similar interface blocks and functions can be written to make the other operations utilize symbols, such as - and *. The precedence of the extended operators when used to compute with big integers is the same as when they are used to add ordinary integers. This holds true for all built-in operators that are extended. The following program tests the extended multiplication operator (the function is not shown). By looking at the last digit of the answer, it is possible to see that the multiplication is done before the addition.

```
program test_big_4

   use big_integers_module
   type(big_integer) :: a, b, c

   a = "1"
   b = "9999999999999999999"
   c = "9999999999999999999"
   print *, a + b * c

end program test_big_4
```

```
99999999999999999980000000000000000002
```

New Operators for Big Integers

In addition to extending the meaning of the built-in operators, it is possible to make up new names for operators. For example, we could define a new operator `.prime.`, whose operand is a big integer and whose value is true if the big integer is a prime and is false otherwise. Its name is used in an interface statement and the function.

```
public :: operator (.prime.)
interface operator (.prime.)
   procedure prime
end interface operator (.prime.)
```

This operator could now be used just like any built-in unary operator, as illustrated by the following `if` statement:

```
if (.prime. b) call print_big(b)
```

Raising a Big Integer to an Integer Power

Exponentiation, like the factorial function, has both an iterative definition and a recursive definition. They are

$$x^n = x \times x \times \cdots \times x \quad n \text{ times}$$

and

$$x^0 = 1$$

$$x^n = x \times x^{n-1} \quad \text{for } n > 1$$

Since Fortran has an exponentiation operator ** for integer and real numbers, it is not necessary to write a procedure to do that. However, it may be necessary to write an exponentiation procedure for a new data type, such as our big integers. We suppose that the multiply operator (*) has been extended to form the product of two big integers.

The task is to write a procedure for the module that will raise a big integer to a power that is an ordinary nonnegative integer. This time, the simple iterative procedure is presented first.

```
function big_power_int(b, i) result(big_power_int_result)

   type(big_integer), intent(in) :: b
   integer, intent(in) :: i
   type(big_integer) :: big_power_int_result
   integer :: n

   big_power_int_result = "1"
   do n = 1, i
      big_power_int_result = big_power_int_result * b
   end do

end function big_power_int
```

It would be straightforward to use the recursive factorial function as a model and construct a recursive version of the exponentiation function; but this is another example of tail recursion, and there is no real advantage to the recursive version. However, think about how you would calculate x^{18} on a calculator that does not have exponentiation as a built-in operator. The clever way is to compute x^2 by squaring x, x^4 by squaring x^2, x^8 by squaring x^4, x^{16} by squaring x^8, and finally x^{18} by multiplying the results obtained for x^{16} and x^2. This involves a lot fewer multiplications than doing the computation the hard way by multiplying x by itself 18 times. To utilize this scheme to construct a program is fairly tricky. It involves computing all the appropriate powers x^2, x^4, x^8, ..., then multiplying together the powers that have a 1 in the appropriate position in the binary representation of n. For example, since $18 = 10010_2$, powers that need to be multiplied are 16 and 2.

It happens that there is a recursive way of doing this that is quite easy to program. It relies on the fact that x^n can be defined with the following less-obvious recursive definition below. The trick that leads to the more efficient recursive exponentiation function is to think of the problem "top-down" instead of "bottom-up". That is, solve the problem of computing x^{18} by computing x^9 and squaring the result. Computing x^9 is almost as simple: square x^4 and multiply the result by x. Eventually, this leads to the problem of computing x^0, which is 1. The recursive definition we are looking for is

$$x^0 = 1$$

$$x^n = \begin{cases} (x^{\lfloor n/2 \rfloor})^2 \\ (x^{\lfloor n/2 \rfloor})^2 \times x \end{cases}$$

where the first line of the second equation is used for n positive and even and the second when n is positive and odd and where $\lfloor\ \rfloor$ is the floor function, which for positive integers is the largest integer less than or equal to its argument. This definition can be used to construct a `big_power_int` function that is more efficient than the iterative version.

```
pure recursive function big_power_int(b, i) &
      result(big_power_int_result)

   type(big_integer), intent(in) :: b
   integer, intent(in) :: i
   type(big_integer) :: big_power_int_result
   type(big_integer) :: temp_big

   ! Code to handle Inf and NaNs omitted
   if (i <= 0) then
      big_power_int_result = "1"
   else
      temp_big = big_power_int(b, i / 2)
      if (modulo(i, 2) == 0) then
         big_power_int_result = temp_big * temp_big
      else
         big_power_int_result = temp_big * temp_big * b
      end if
   end if

end function big_power_int
```

Exercises

1. Extend the equality operator (==) and the "less than" (<) operator to compare two big integers.
2. Extend the equality operator (==) to compare a big integer with a character string consisting of digits. *Hint:* Use extended assignment to assign the character string to a temporary big integer, then use the extended equality operator from Exercise 1 to do the comparison.
3. Extend the multiplication operator (*) to two big integers.
4. Use the result of the previous exercise to compute $100! = 100 \times 99 \times \cdots \times 2 \times 1$. It may be necessary to increase the value of the parameter `nr_of_digits`.
5. Extend the subtraction operator (-) so that it performs "positive" subtraction. If the difference is negative, the result should be 0.
6. Extend the intrinsic function `huge` to apply to a big integer argument.

7. Extend the dt subroutine print_point so that if an explicit dt edit descriptor is used with the character asterisk (so that DT* is passed to the dummy argument iotype), an object of type point is printed as if the format were list directed (*).

8. Extend the dt subroutine print_big so that one integer value *w* can be passed to v_list. If *w* is passed (that is, v_list is not zero sized), have the dt edit descriptor behave just like the I*w* edit descriptor.

9. The representation of big integers used in this section is very inefficient because only one decimal digit is stored in each Fortran integer array element. It is possible to store a number as large as possible, but not so large that when two are multiplied, there is overflow. This largest value can be determined portably on any system with the statements

```
integer, parameter :: &
      d = (range(0) - 1) / 2, &
      base = 10 ** d

! Base of number system is 10 ** d,
! so that each "digit" is 0 to 10**d - 1
```

   On a typical system that uses 32 bits to store an integer, with 1 bit used for the sign, the value of the intrinsic inquiry function range(0) is 9 because $10^9 < 2^{31} < 10^{10}$. To ensure that there is no chance of overflow in multiplication, this number is decreased by one before dividing by two to determine the number of decimal digits d that can be stored in one array element digit of a big integer. In our example, this would set d to 4. The value of base is then 10**d, or $10^4 = 10{,}000$. With this scheme, instead of storing a number from 0 to 9 in one integer array element, it is possible to store a number from 0 to base – 1, which is 9,999 in the example. In effect, the big number system uses base 10,000 instead of base 10 (decimal).

   Determine the value of range (0) on your system.

10. Modify the type definition for big_integer so that a number from 0 to base – 1 (previous exercise) is stored in each element of the array. The number of elements in the array should be computed from the parameter nr_of_digits.

11. Determine the largest number that can be represented as the value of a big integer using the type definition in the previous exercise.

12. Modify the procedure big_gets_char to use the more efficient representation of big integers.

13. Modify the procedure print_big to use the more efficient representation of big integers. In the format, i1 should be replaced by i*d.d*, where *d* is the number of decimal digits stored in each array element.

14. Modify the subroutine `big_plus_big` using the new type definition for `big_integer`. It is very similar to the one developed in this section, except that the base is now not 10, but `base`.

15. Extend the operator * to multiply a big integer by an ordinary integer.

16. Extend `huge` using the new representation. Write a test program that prints `huge(b)`.

17. Approximately n multiplications are required to compute x^n by the iterative version of the function `big_power_int`. Estimate the number of multiplications needed to compute x^n by the recursive version.

18. *Project:* Write a module to do computation with rational numbers. The rational numbers should be represented as a structure with two integers, the numerator and the denominator. Provide assignment, some input/output, and some of the usual arithmetic operators. Addition and subtraction are harder than multiplication and division, and equality is nontrivial if the rational numbers are not reduced to lowest terms.

19. Modify the module in the previous exercise to use big integers for the numerator and denominator.

20. *Project:* Write a module to manipulate big decimal numbers such as

```
28447305830139375750302.37429125612092391236362 92
```

using the `big_integer` module as a model.

Pointer and Allocatable Variables 10

In Fortran, a **pointer variable** or simply a **pointer** is best thought of as a "free-floating" name that may be associated dynamically with or "aliased to" some object. The object already may have one or more other names or it may be an unnamed object. The object may represent data (a variable, for example) or be a procedure.

Syntactically, a pointer is just any sort of variable that has been given the pointer attribute in a declaration. A variable with the pointer attribute may be used just like any ordinary variable, but it may be used in some additional ways as well. To understand how Fortran pointers work, it is almost always better to think of them simply as aliases. Another possibility is to think of the pointers as "descriptors", sufficient to describe a row of a matrix, for example.

Allocatable arrays were discussed in 4.1; in this chapter, allocatable scalars are discussed.

Both pointers and allocatable arrays can be used to create recursive structures, such as linked lists, queues, and trees.

In many applications, either pointers or allocatables could be used. When possible, it is usually preferable to use allocatable variables as they are less error-prone, more efficient, and will not cause memory leaks.

10.1 The Use of Pointers in Fortran

Each pointer in a Fortran program is in one of the three following states:

1. It may be **undefined**, which is the condition of each pointer at the beginning of a program, unless it has been initialized.

2. It may be **null**, which means that it is not the alias of any data object.

3. It may be **associated**, which means that it is the alias of some target data object.

The terms "disassociated" and "not associated" are used when a pointer is in state 2. A pointer is defined if it is in state 2 or 3. The `associated` intrinsic inquiry function discussed later distinguishes between states 2 and 3 only; its arguments must be defined.

The Pointer Assignment Statement

To start with a very simple example, suppose p is a real variable with the pointer attribute, perhaps given with the declaration

```
real, pointer :: p
```

Suppose r is also a real variable. Then it is possible to make p an alias of r by the **pointer assignment statement**

```
p => r
```

For those that like to think of pointers, rather than aliases, this statement causes p to point to r. Any variable aliased or "pointed to" by a pointer must be given the **target attribute** when declared, and the target must have the same type (class if polymorphic (12.2)), kind, and rank as the pointer. If the object is polymorphic, the pointer assumes the type of the object. However, it is not necessary that the variable has a defined value. For procedures, the interfaces of the pointer and target must be the same. For our example above, these requirements are met by the presence of the following declaration:

```
real, target :: r
```

A variable with the pointer attribute may be an object more complicated than a simple variable. It may be an array or structure, for example. The following declares v to be a pointer to a one-dimensional array of reals:

```
real, dimension(:), pointer :: v
```

With v so declared, it may be used to alias any one-dimensional array of reals, including a row or column of some two-dimensional array of reals. For example,

```
v => real_array(4, :)
```

makes v an alias of the fourth row of the array real_array. Of course, real_array must have the target attribute for this to be legal.

```
real, dimension(100, 100), target :: real_array
```

Once a variable with the pointer attribute is an alias for some data object, that is, it is pointing to something, it may be used in the same way that any other variable with the same properties may be used. For the example above using v,

```
print *, v
```

has exactly the same effect as

```
print *, real_array(4, :)
```

and the assignment statement

```
v = 0
```

has the effect of setting all the elements of the fourth row of the array `real_array` to 0.

An array pointer on the left-hand side of a pointer assignment may specify lower bounds.

```
real, dimension(:, :), pointer :: p
real, dimension(100, -100:100), target :: t
   . . .
p(0:, 1:) => t
```

The bounds of p are 0:99 and 1:201.

A different version of the pointer assignment statement occurs when the right side also is a pointer. This is illustrated by the following example, in which p1 and p2 are both real variables with the pointer attribute and r is a real variable with the target attribute.

```
real, target :: r
real, pointer :: p1, p2
r = 4.7
p1 => r
p2 => p1
r = 7.4
print *, p2
```

After execution of the first assignment statement, r is a name that refers to the value 4.7:

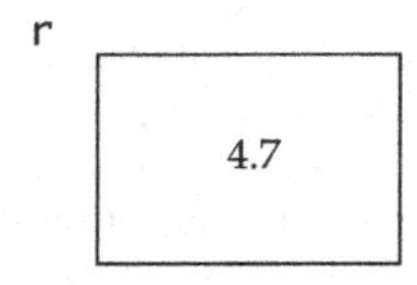

The first pointer assignment causes p1 to be an alias for r, so that the value of the variable p1 is 4.7. The value 4.7 now has two names, r and p1, by which it may be referenced.

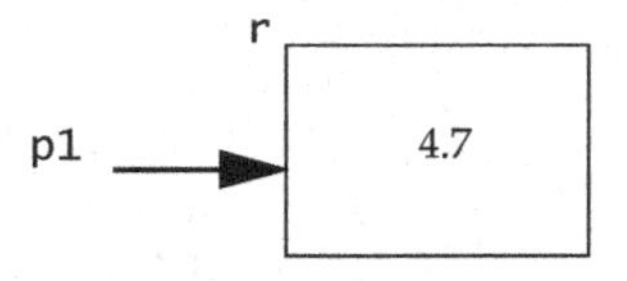

The next pointer assignment

```
p2 => p1
```

causes p2 to be an alias for the same thing that p1 is an alias for, so the value of the variable p2 is also 4.7. The value 4.7 now has three names or aliases, r, p1, and p2.

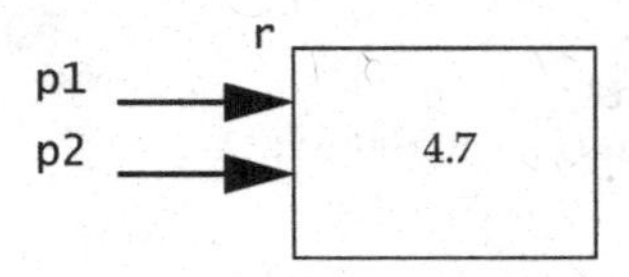

Changing the value of r to 7.4 causes the value of both p1 and p2 also to change to 7.4 because they are both aliases of r. Thus, the next print statement

```
print *, p2
```

prints the value 7.4.

The pointer assignment statement

```
p => q
```

is legal whatever the status of q. If q is undefined, p is undefined; if it is null, p is nullified; and if it is aliased to or associated with a target, p becomes associated with the same target. Note that if q is associated with some target, say t, it is not necessary that t has a defined value.

The Difference between Pointer and Ordinary Assignment

We can now illustrate the difference between pointer assignment, which transfers the status of one pointer to another, and ordinary assignment involving pointers. In an ordinary assignment in which pointers occur, the pointers must be viewed simply as aliases for their targets. Consider the following statements:

```
real, pointer :: p1, p2
real, target  :: r1, r2
   . . .
r1 = 1.1
r2 = 2.2
p1 => r1
p2 => r2
```

This produces the following situation:

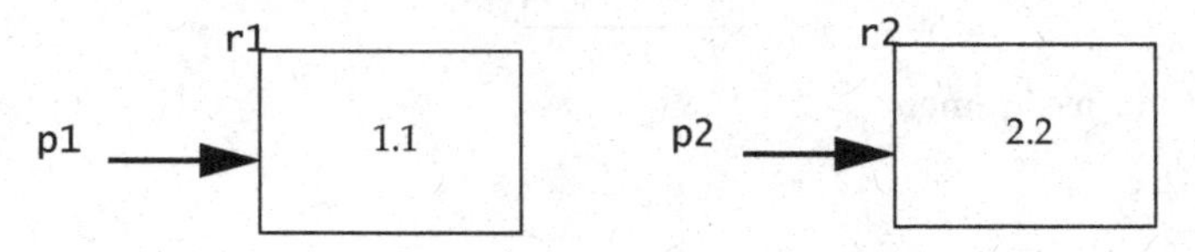

Now suppose the ordinary assignment statement

```
p2 = p1
```

is executed. This statement has exactly the same effect as the statement

```
r2 = r1
```

because p2 is an alias for r2 and p1 is an alias for r1. The situation is now:

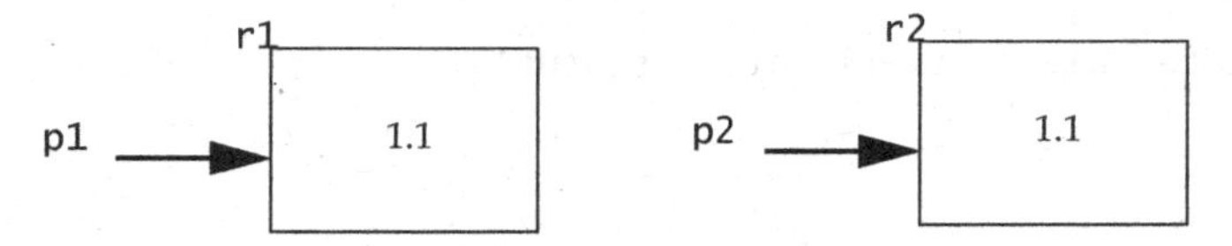

because the value 1.1 has been copied from r1 to r2. The values of p1, p2, r1, and r2 are all 1.1. Subsequent changes to r1 or p1 will have no effect on the value of r2.

If, on the other hand, the pointer assignment statement

```
p2 => p1
```

were executed instead, this statement would produce the situation

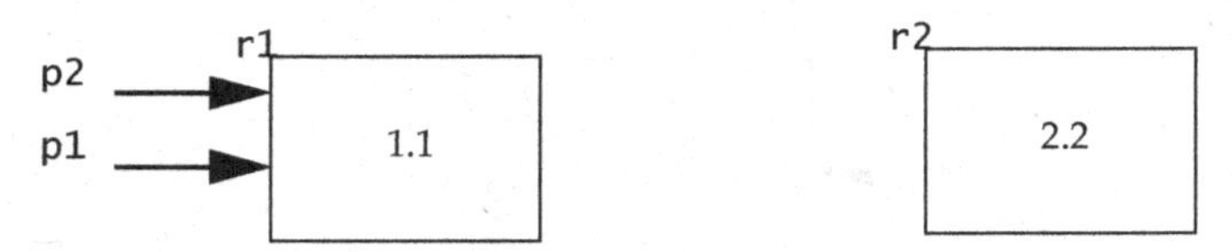

In this case, too, the values of p1, p2, and r1 are 1.1, but the value of r2 remains 2.2. Subsequent changes to p1 or r1 do change the value of p2. They do not change the value of r2.

If the target of p1 is changed to r2 by the pointer assignment statement

```
p1 => r2
```

the target r1 and value 1.1 of p2 do not change, producing the following situation:

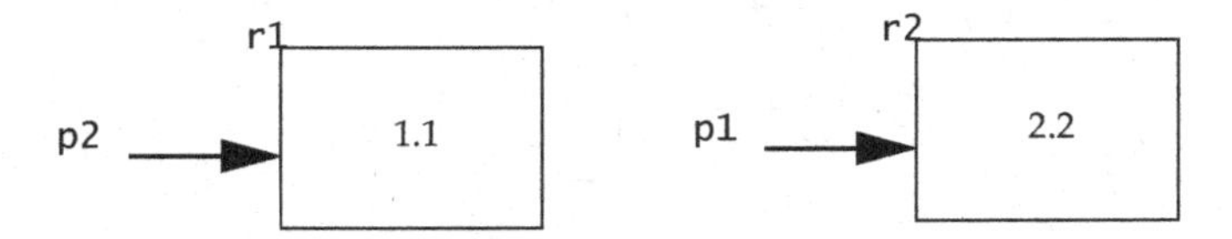

The pointer p2 remains an alias for r1; it does not remain associated with p1.

The allocate and deallocate Statements for Pointers

With the **allocate statement**, it is possible to create space for a value and cause a pointer variable to refer to that space. The space has no name other than the pointer mentioned in the allocate statement. For example,

```
allocate (p1)
```

creates space for one real number and makes `p1` an alias for that space. No real value is stored in the space by the `allocate` statement, so it is necessary to assign a value to `p1` before it can be used, just as with any other real variable.

As in the `allocate` statement for allocatable arrays, it is possible to test if the allocation is successful. This might be done with the statement

```
allocate (p1, stat=allocation_status)
```

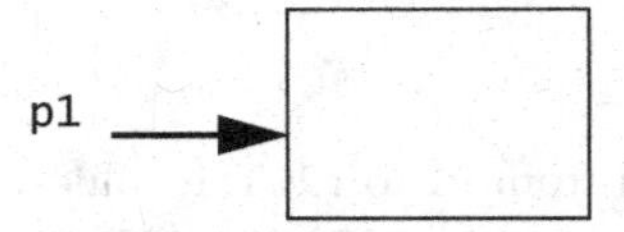

The statement

```
p1 = 7.7
```

sets up the following situation.

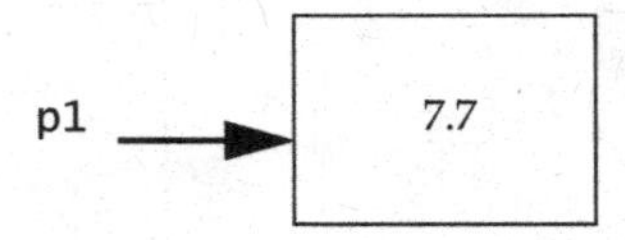

Before a value is assigned to `p1`, it must either be associated with an unnamed target by an `allocate` statement or be aliased with a target by a pointer assignment statement.

The `deallocate` statement throws away the space pointed to by its argument and makes its argument null (state 2). For example,

```
deallocate (p1)
```

disassociates `p1` from any target and nullifies it.

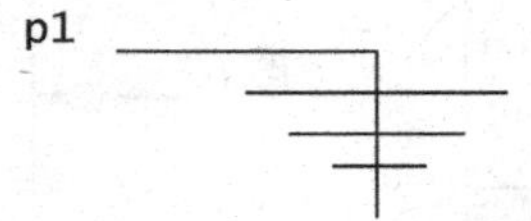

After `p1` is deallocated, it must not be referenced in any situation that requires a value; however it may be used, for example, on the right side of a pointer assignment statement. If other pointer variables were aliases for `p1`, they, too, no longer reference a value; however, they are not nullified automatically.

The `null` Intrinsic Function

At the beginning of a program, a pointer variable (just as all other variables) is not defined, unless it is initialized. A pointer variable must not be referenced to produce a value when it is not defined, but it is sometimes desirable to have a pointer variable be in the state of not pointing to anything, which might signify the last item in a linked list, for example. This occurs when it is set to the value of the **null** intrinsic function, which creates a condition that may be tested and assigned to other pointers by pointer assignment (=>). A pointer is nullified with a pointer assignment such as

```
p1 => null()
```

If the target of p1 and p2 are the same, nullifying p1 does not nullify p2. On the other hand, if p1 is null, then executing the pointer assignment

```
p2 => p1
```

causes p2 to be null also.

A null pointer is not associated with any target or other pointer.

The `associated` Intrinsic Function

The **associated** intrinsic function may be used to determine if a pointer variable is pointing to, or is an alias for, another object. To use this function, the pointer variable must be defined; that is, it must either be the alias of some data object or be null. The associated function indicates which of these two cases is true; thus it provides the means of testing if a pointer is null.

The associated function may have a second argument. If the second argument is a target, the value of the function indicates whether the first argument is an alias of the second argument. If the second argument is a pointer, it must be defined; in this case, the value of the function is true if both pointers are null or if they are both aliases of the same target. For example, the expression

```
associated(p1, r)
```

indicates whether or not p1 is an alias of r, and the expression

```
associated(p1, p2)
```

indicates whether p1 and p2 are both aliases of the same thing or they are both null.

If two pointers are aliases of different parts of the same array, they are not considered to be associated. For example, the following program will print the value false.

```
program test_associated
   implicit none
   real, target, dimension(4) :: a = [ 1, 2, 3, 4 ]
   real, pointer, dimension(:) :: p, q
   p => a(1:3)
   q => a(2:4)
```

```
   print *, associated(p, q)
end program test_associated
```

Pointer Remapping

With pointer remapping, a one-dimensional pointer may alias a part of another array that it could not otherwise. An example is the diagonal of a matrix.

```
real, pointer, dimension(:, :) :: matrix
real, pointer, dimension(:) :: diagonal, base
   . . .
allocate (base(n*n))
matrix(1:n, 1:n) => base
diagonal => base(::n+1)
```

`diagonal` now points to the diagonal elements of `matrix`.

Procedure Pointers

A procedure may have the `pointer` attribute. The procedure may not be generic or elemental.

When a procedure pointer is declared, the characteristics of procedures it may alias are given by referencing a procedure interface (3.9); the interface may be abstract, the interface of an actual procedure, or the name of a handy procedure.

```
procedure (simple_sub), pointer :: sss
procedure (f), pointer :: fff => null()
   . . .
sss => subr
call sss()  ! calls subr
   . . .
fff => cos
print *, fff(0.24)  ! prints cos(0.24)
```

Arrays of Procedures

Derived types and procedure pointers can be used to create, in effect, an array of procedures. The trick is to create an array of structures, each containing one component that is a procedure pointer.

```
type, public :: proc_type
   procedure (real_f_x), pointer, nopass :: &
      ptr_to_f => null()
end type proc_type
type (proc_type), dimension(:), allocatable :: ap
```

Without the `nopass` attribute (12.3), it is assumed that the first argument of `ptr_to_f` would be an object of type `proc_type`.

```
intrinsic :: sin
   . . .
allocate (ap(100))
ap(1)%ptr_to_f => sin
   . . .
do n = 1, size(ap)
   print *, ap(n)%ptr_to_f(x)
end do
```

Dangling Pointers and Unreferenced Storage

There are two situations that the Fortran programmer must avoid. The first is a **dangling pointer**. This situation arises when a pointer variable is an alias for some object that gets deallocated by an action that does not involve the pointer directly. For example, if `p1` and `p2` are aliases for the same object, and the statement

```
deallocate (p2)
```

is executed, it is obvious that `p2` is now disassociated, but the status of `p1` appears to be unaffected, even though the object to which it was pointing has disappeared. A reference to `p1` is now illegal and will produce unpredictable results. It is the responsibility of the programmer to keep track of the number of pointer variables referencing a particular object and to nullify each of the pointers whenever one of them is deallocated.

A related problem of **unreferenced storage** can occur when a pointer variable that is an alias of an object is nullified or set to alias something else without a deallocation. If there is no other alias for this value, it is still stored in memory somewhere, but there is no way to refer to it. This is not important if it happens to a few simple values, but if it happens many times to large arrays, the efficient management of storage could be hampered severely. In this case, it is also the responsibility of the programmer to ensure that objects are deallocated before all aliases of the object are modified. Fortran systems are not required to have runtime "garbage collection" to recover the unreferenced storage, but some do.

Exercise

1. Construct an array of three structures, each of which has a procedure pointer as its only component (as shown in this section). Set the three pointers to alias the intrinsic functions `sin`, `cos`, and `tan`.

   Create a table of trig functions with the sin(x), cos(x), and tan(x) on one line and a line each for x = 0.1, 0.2, ..., 1.0. Print out one or two values from the table to make sure they are correct.

10.2 Moving Pointers vs. Moving Data

In some cases, pointers can be used to avoid copying large amounts of data. In the heat transfer program in 4.6, each time new temperature values are temporarily stored so that they may be compared with the old values, then copied back into the interior of the plate.

```
temp = (n + e + s + w) / 4.0
   . . .
inside = temp
```

If a second plate temp_plate (temporary plate) is used, then it is possible to store the results of the calculation in the interior of the temporary plate. Then after convergence is tested, the pointer plate can be set to temp_plate, avoiding a data copy. A temporary pointer temp must be used to save the value of plate, so it can be deallocated; otherwise, after many iterations, there would be many copies of unreferenced storage.

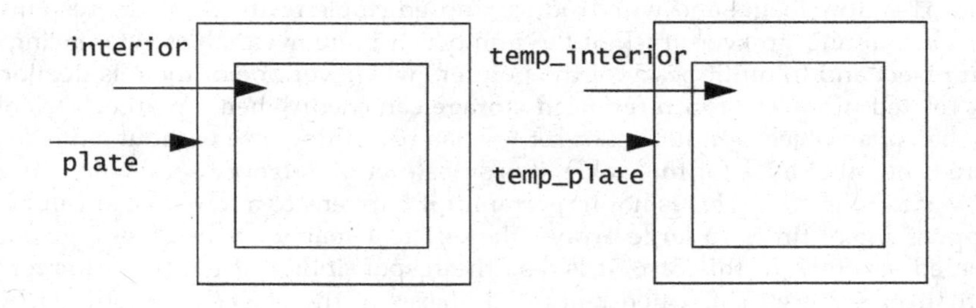

Figure 10-1 Exchanging pointers to "move" data

Here is part of the code

```
temp_interior = (n + e + s + w)/4.0
diff = maxval(abs(temp_interior-inside))

! Exchange pointers
temp => plate
plate => temp_plate
temp_plate => temp
```

For this example, the performance improvement was only about 10–20%, depending on the size P of the plate because the time to compute the new values is larger than the time to move the data.

10.3 Linked Lists

Linked lists have many uses in a wide variety of application areas; one example in science and engineering is the use of a linked list to represent a queue in a simulation program. Lists of values can be implemented in Fortran in more than one way. Perhaps the most obvious way is to use an array. Another is to use pointers and data structures to create a linked list. The choice should depend on which operations are going to be performed on the list and the relative frequency of those operations. If the only requirement is to add and delete numbers at one end of the list, as is done if the list is treated as a stack, then an array is an easy and efficient way to represent the list. If items must be inserted and deleted often at arbitrary points within the list, then a linked list is nice; with an array, many elements would have to be moved to insert or delete an element in the middle of the list. Another issue is whether storage is to be allocated all at once, using an array, or element by element in a linked list implementation. The implementation of linked lists using pointers also uses recursion effectively, but iteration also could be used.

A **linked list** of numbers (or any other objects) can be thought of schematically as a bunch of boxes, often called *nodes,* each containing a number and *pointer* to the box containing the first number in the rest of the list. Suppose, for example, the list contains the numbers 14, 62, and 83. In the lists discussed in this section, the numbers are kept in numeric order, as they are in this example. Figure 10-2 contains a pictorial representation of the list.

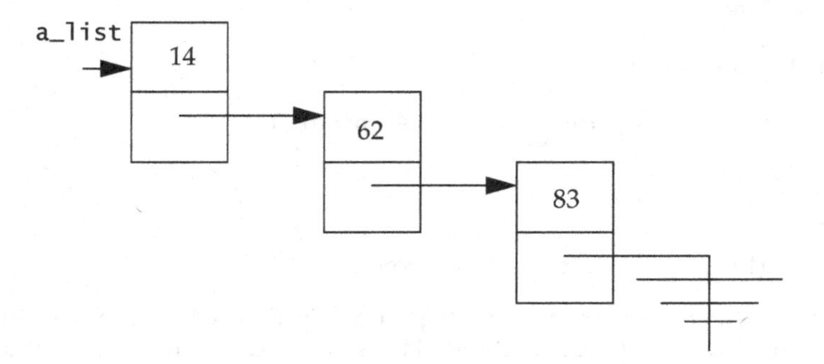

Figure 10-2 A linked list of integers

We will illustrate the Fortran techniques for manipulating linked lists by constructing a module to manipulate linked lists in which the nodes contain integers and the lists are sorted with the smallest number at the head of the list. The procedures of the module `sorted_integer_lists_module` use recursion. The recursion is usually tail recursion, so, in one sense, not much is gained. However, it turns out that much of the detailed manipulation of pointers is eliminated and the recursive versions do not need such things as a dummy node at the head of the list or "trailing pointers", which are necessary in the nonrecursive implementations of linked lists. This makes the routines a lot easier to write, understand, and maintain, but perhaps a little less efficient to execute on systems that have a high overhead for procedure calls.

Recursive List Processing

The approach to writing recursive routines to process a list is to view the list itself as a recursive data structure. That is, a list of integers is either empty or it is an integer followed by a list of integers. This suggests that to process a list of numbers, process the first number in the list and then process the rest of the list with the same routine, quitting when the list is empty.

To view the list as a recursive data structure as described above, a `node` should consist of a value and another object, `next`, of type `sorted_list`. An object of type `sorted_list` is a pointer. The overall structure of the module and the type definitions for the module are

```
module sorted_integer_lists_module

   implicit none
   private

   type, public :: sorted_list
      private
      integer :: value
      type(sorted_list), pointer :: next => null()
   contains
      final :: empty
   end type sorted_list

public:: is_empty, insert, delete, print_list

contains
   . . .

end module sorted_integer_lists_module
```

The `public` statement lists the module procedures that will be available to any program using the module `sorted_integer_lists_module`. The `public` attribute on the definition of the type `list` indicates that the type will also be available to programs that use the module. Thus, the user is able to declare variables to be type `sorted_list` and process lists with the public module procedures `insert`, `empty`, `delete`, and `print_list`. These could be type-bound procedures if a more object-oriented (12) approach were desired.

The `private` statement in the definition of the type `sorted_list` indicates that although the type `sorted_list` is available to any program that uses the module, the user will not be able to access the internal structure. The component `next` is default initialized to null, which means that each time an object of type `list` is declared or allocated, the component `next` is null.

The type `sorted_list` has a final subroutine named `empty`. This means that the user of a module can simply deallocate any sorted list and all the parts of the linked list will be deallocated, returning their storage to the pool of available storage. It is not listed in the `public` statement, because a final subroutine is always available.

Abstract Data Types

The purpose is to provide the user of the module with an **abstract data type**, that is, with the name of the list type and all necessary procedures to manipulate these lists. If it is desirable to change the implementation, we can be sure that no program has accessed the lists in any way except those provided by the public procedures in this module.

It is important to be able to declare which details of a module are private to the module and therefore hidden from all external users of the module, and which are public.

Now we must supply procedures for the operations that are needed. They are the function `is_empty(list)` that returns the logical value indicating whether or not the list is empty; a subroutine `insert(list, number)` that inserts `number` a number into a list; a subroutine `delete(list, number, found)` that deletes one occurrence of a number from a list, if it is there, and indicates if it is found; and a subroutine `print_list (list)` that prints the numbers in the list in order. Some of these are quite simple and could be done easily without a procedure, but the purpose is to include all necessary operations in the module `sorted_integer_lists_module` and be able to change the implementation.

Inserting a Number

Let us first do the subroutine that inserts a number into a list. The recursive version is deceptively simple. First, if the list is empty, another list must be created and the number placed in its `value` field. Then the `next` field is made empty because there are no other elements in the list. The Fortran statements to do this are

```
allocate (list)
list%value = number
```

If the list is not empty, the number to be inserted must be compared with the first value in the list. If it is smaller or equal, it must be inserted as the first value of the list. Again, a new first node is created and the number placed in its `value` field. However, this time the `next` field of this new first node is set equal to the original list before the insertion because all other numbers in the list follow this new number. The Fortran statements to do this are

```
allocate (temp)
temp = sorted_list(number, list)
list => temp
```

A temporary variable of type `list` is necessary to prevent losing the reference to the list when a new first node is allocated. Notice that pointer assignment (=>) is used for the variables of type `sorted_list`.

These are the nonrecursive base cases of the recursive subroutine `insert`. The only other remaining case is when the number to be inserted is greater than the first element of the list. In this case, the insertion is completed by a recursive call to insert the number in the rest of the list. The complete subroutine `insert` follows.

```
recursive subroutine insert(list, number)

   type(sorted_list), pointer, intent(in out) :: list
   integer, intent(in) :: number
   type(sorted_list), pointer :: temp

   if (is_empty(list)) then
      allocate (list)
      list%value = number
   else if (number <= list%value) then
      temp => list
      allocate (list)
      list = sorted_list(number, temp)
   else
      call insert(list%next, number)
   end if

end subroutine insert
```

As is typical with recursive algorithms, the program listing appears simpler than the execution. To help understand why the recursive subroutine `insert` works, we simulate its execution to insert the number 62 in a sorted list containing 14 and 83. The list supplied to the dummy argument `list` at the top level call to `insert` is shown in Figure 10-3. The first item in the list is less than 62, so a second level call to `insert` is

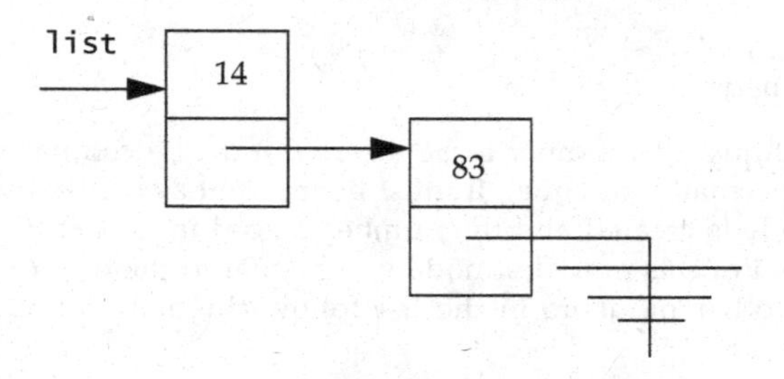

Figure 10-3 Top level call to insertion subroutine

made to insert 62 into the rest of the list as shown in Figure 10-4.

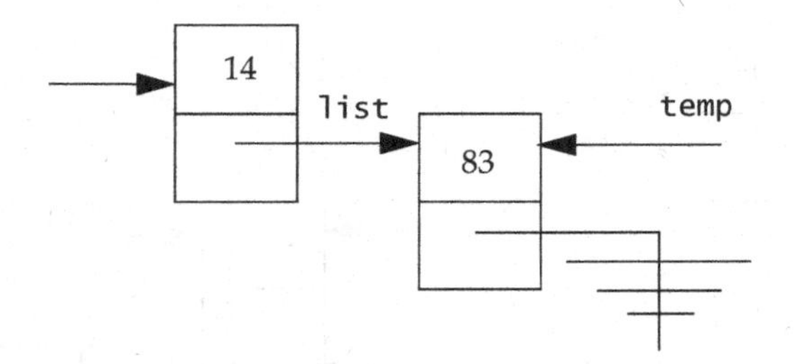

Figure 10-4 Second level call to insertion subroutine

This time the new number 62 is less than the first element 83 of the list referenced by dummy argument `list`, so the nonrecursive alternative to insert before the first element is selected. Figure 10-5 shows the situation after the `allocate` statement.

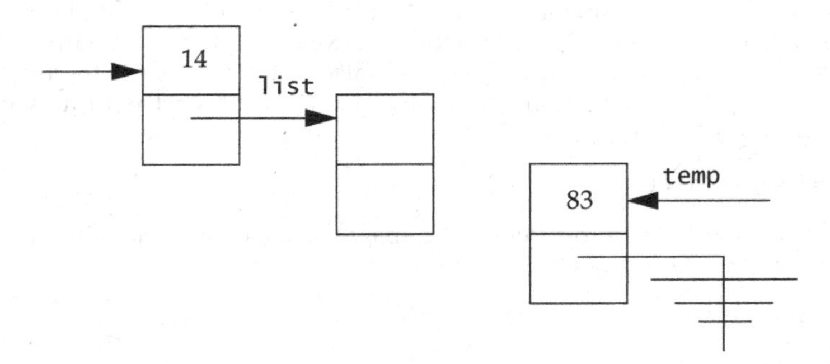

Figure 10-5 Allocating a new node for a linked list

Figure 10-6 shows the situation after the structure assignment statement that links the new node to the rest of the list starting with the number 83.

Determining if a List is Empty

The function that determines if a list is empty is straightforward. Recall that a pointer is not associated if it is null.

```
function is_empty(list) result(is_empty_result)

   type(sorted_list), pointer, intent(in) :: list
   logical :: is_empty_result
   is_empty_result = .not. associated(list)

end function is_empty
```

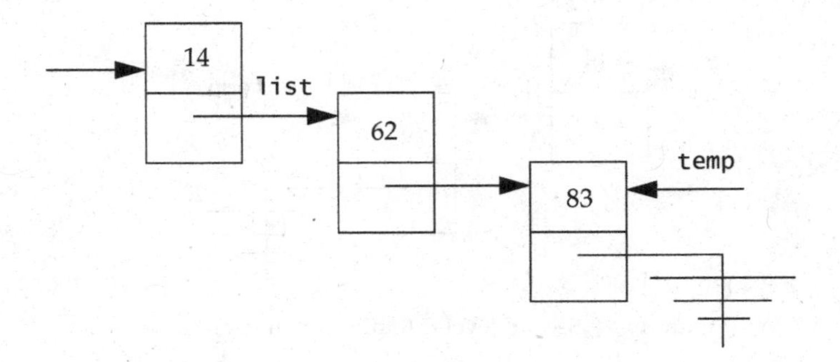

Figure 10-6 The linked list after the new number is inserted

Deleting a Number

The subroutine to delete a number from a list, if it is there, is quite similar to the subroutine to insert. There are two special nonrecursive cases. If the list is empty, the number cannot be deleted from it, so `found` is set to false. If the number is the first number in the list, deleting it may be accomplished by making `list` start with its second element (if any) using the statement

```
list => list%next
```

Also, it is a good idea to deallocate the space for the deleted node to avoid unreferenced storage. The statements

```
temp => list

temp%next => null()
deallocate (temp)
```

accomplish this. The first of these must be done before `list` is reassigned, and the others afterward. `temp%next` must be set to null, so that when `temp` is deallocated, the final subroutine `empty` does not deallocate the entire remainder of the list.

In case the list is not empty, but the desired number is not its first element, the number is deleted by a recursive call to delete it from the rest of `list`. The full subroutine `delete` follows.

```
recursive subroutine delete(list, number, found)

   type(sorted_list), pointer, intent(in out) :: list
   integer, intent(in) :: number
   logical, intent(out) :: found
   type(sorted_list), pointer :: temp
```

```
   if (is_empty(list)) then
      found = .false.
   else if (list%value == number) then
      ! Delete node pointed to by list
      temp => list
      list => list%next
      temp%next => null()
      deallocate(temp)
      found = .true.
   else
      call delete(list%next, number, found)
   end if

end subroutine delete
```

For example, if the number 62 is to be deleted from a list with elements 14, 62, and 83, the first call has `list` pointing to the node containing 14. Because this is not the desired number, a second call is made with `list` pointing to the node containing 62, as shown in Figure 10-7. This node is deleted by making the `next` field of the node containing 14 (which is the actual argument corresponding to the dummy argument `list`) point to the node containing 83 and deallocating the node containing 62.

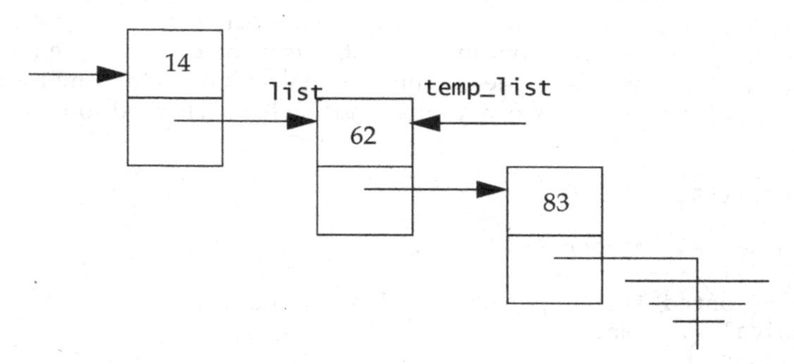

Figure 10-7 The second (recursive) call to the deletion routine

The subroutine `print_list` prints the numbers in the list in order. This just involves recursively traversing the list as in the subroutines `insert` and `delete`.

```
recursive subroutine print_list(list)

   type(sorted_list), pointer, intent(in) :: list

   if (associated(list)) then
      write (unit=*, fmt="(tr1, i0)", advance="no") list%value
      call print_list(list%next)
   end if

end subroutine print_list
```

Although this is just an instance of tail recursion, the procedure is simpler than any iterative version.

The Final Subroutine `delete`

Whenever a list of type `sorted_list` is deallocated, the final subroutine `empty` is called with the list as its argument. The subroutine deletes the node pointed to by the argument `list` and recursively deletes the list referenced by its `next` component.

```
recursive subroutine empty(list)

   type(sorted_list), intent(in out) :: list
   if (associated(list%next)) then
      deallocate (list%next)
   end if

end subroutine empty
```

Sorting with a Linked List

With the integer list module just created, it is possible to write a simple but inefficient sorting program. The program works by inserting numbers into a list that is ordered. When all the numbers have been put into the list, it is printed, producing all the numbers in order. Just to show that the delete subroutine works the odd numbers are deleted from the list. Finally the list is deallocated and the list is checked to make sure it is empty.

```
program list_sort

   use sorted_integer_lists_module
   implicit none
   type(sorted_list), pointer :: list => null()
   logical :: found
   integer :: n
   integer, dimension(*), parameter :: numbers = &
      [ 4, 6, 3, 8, 7, 9, 2, 1, 5 ]

   do n = 1, size(numbers)
      call insert(list, numbers(n))
   end do

   print *, "Sorted list"
   call print_list(list)

   do n = 1, size(numbers)
      if (modulo(numbers(n), 2) /= 0) then
         call delete(list, numbers(n), found)
```

```
         if (.not. found) then
            print *, numbers(n), "not found in list"
         end if
      end if
   end do

   print *; print *
   print *, "List with odd numbers deleted"
   call print_list(list)
   deallocate(list)
   print *; print *
   print *, "Is list empty?", is_empty(list)

end program list_sort
```

Running the program produces the following output.

```
Sorted list
1 2 3 4 5 6 7 8 9

List with odd numbers deleted
2 4 6 8

Is list empty? T
```

Style note: A very important point to note is that even when the procedures in the module `sorted_integer_lists_module` are rewritten to use iteration instead of recursion, or even if arrays are used to represent the lists, creating yet another list module, a program such as `list_sort` that uses one of these modules does not have to be changed at all (unless the name of the module is changed). This illustrates one of the real benefits of using modules. However, although the source code for the program is unchanged, whenever a module changes, any program that uses the module must be recompiled.

Exercise

1. Create a different version of the `sorted_integer_lists_module` with all of the same public types and procedures. However, this version should implement a list with a dynamic array, rather than a linked list.

2. Use first the linked list version and then the array version of the programs to manipulate lists of integers to construct a program that sorts integers. Experiment with each program, sorting different quantities of randomly generated integers to determine an approximate formula for the complexity of the program. Is the execution time (or some other measure of complexity, such as the number of statements executed) proportional to $n \log_2 n$? Is it proportional to n^2?

10.4 Trees

One of the big disadvantages of using a linked list to sort numbers is that the resulting program has poor expected running time. In fact, for the program `list_sort`, the expected running time is proportional to n^2, where n is the number of numbers to be sorted. A much more efficient sorting program can be constructed if a slightly more complicated data structure, the binary tree, is used. The resulting program, `tree_sort`, has an expected running time proportional to $n \log_2 n$ instead of n^2.

It is quite difficult to write nonrecursive programs to process trees, so we will think of trees as recursive structures right from the start. Using this approach, a **binary tree** of integers is either empty or is an integer, followed by two binary trees of integers, called the *left subtree* and *right subtree*.

Sorting with Trees

To sort numbers with a tree, we will construct a special kind of ordered binary tree with the property that the number at the "top" or "root" node of the tree is greater than all the numbers in its left subtree and less than or equal to all the numbers in its right subtree. This partitioning of the tree into a left subtree containing smaller numbers and a right subtree containing larger numbers is exactly analogous to the partitioning of a list into smaller and larger sublists that makes quick sort (4.3) an efficient algorithm. This property will hold not only for the most accessible node at the "top" of the tree (paradoxically called the "root" of the tree), but for all nodes of the tree. To illustrate this kind of tree, suppose a list of integers contains the numbers 265, 113, 467, 264, 907, and 265 in the order given. To build an ordered binary tree containing these numbers, first start with an empty tree. Then place the first number 265 in a node at the root of the tree, as shown in Figure 10-8.

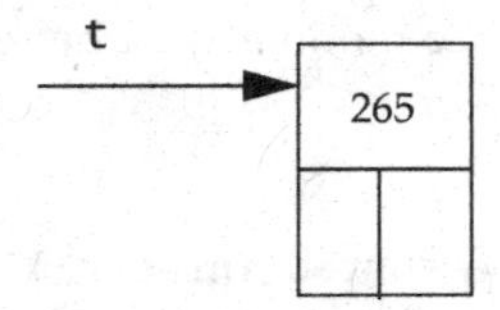

Figure 10-8 The root of a tree

The blank boxes in the lower part of a node are understood to represent unallocated subtrees.

Then the next number is compared with the first. If it is less than the first number, it is placed as a node in the left subtree; if it is greater than or equal to the first number, it is placed in the right subtree. In our example, 113 < 265, so a node containing 113 is created and assigned as the value of the left subtree of the node containing 265, as shown in Figure 10-9.

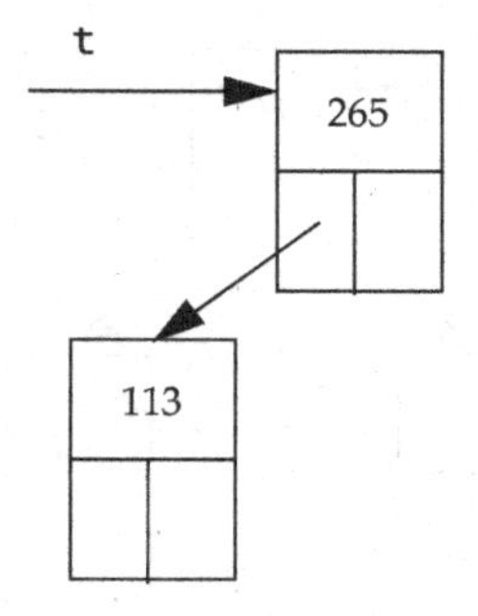

Figure 10-9 Adding the number 113 to the tree

The next number is 467, and it is placed in the right subtree of 265 because it is larger than 265. The result is shown in Figure 10-10.

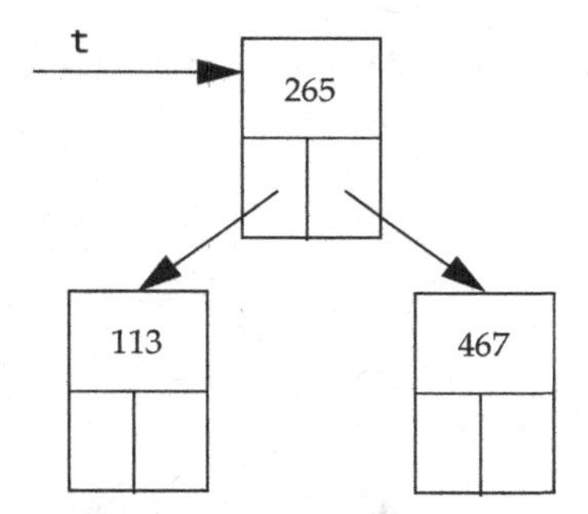

Figure 10-10 Adding the number 467 to the tree

The next number is 264, so it is placed in the left subtree of 265. To place it properly within the left subtree, it is compared with 113, the occupant of the top of the left subtree. Since 264 > 113, it is placed in the right subtree of the one with 113 at the top to obtain the tree shown in Figure 10-11.

The next number 907 is larger than 265, so it is compared with 467 and put in the right subtree of the node containing 467, as shown in Figure 10-12.

The final number 265 is equal to the number in the root node. An insertion position is therefore sought in the right subtree of the root. Since 265 < 467, it is put to the left of 467, as shown in Figure 10-13. Notice that the two nodes containing the number 265 are not even adjacent, nor is the node containing the number 264 adjacent to either node with key 265. This does not matter. When the tree is printed, they will come out in the right order.

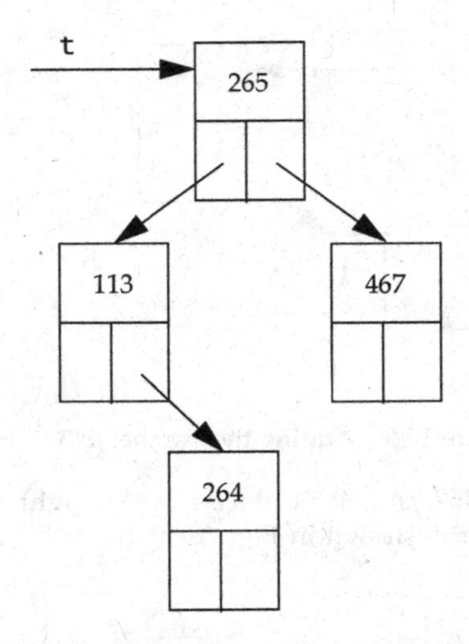

Figure 10-11 Adding the number 264 to the tree

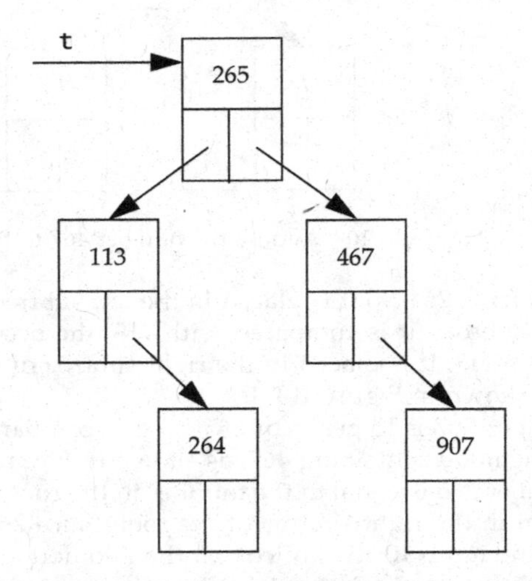

Figure 10-12 Adding the number 907 to the tree

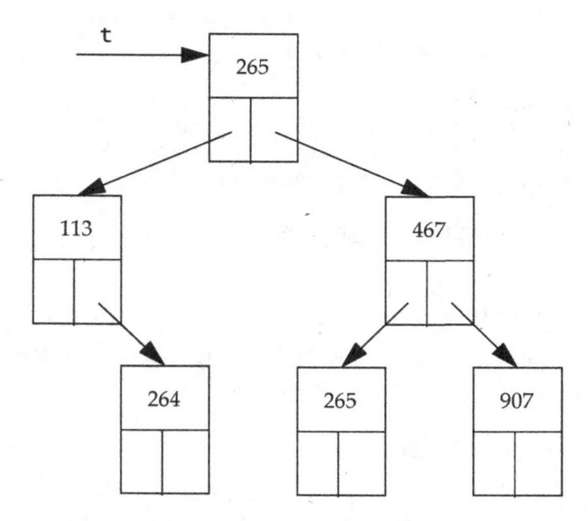

Figure 10-13 The final ordered binary tree

Type Declarations for Trees

The declaration for the node of a tree is similar to the declaration for the node of a linked list, except that the node must contain, in addition to the value, two components, one for the left subtree and one for the right subtree. As with lists, we could have `tree` be a derived data type, which implies it must be a structure with one component, a pointer to the node of a tree. This time, the components will be allocatable scalars instead of pointers. Thus, the declaration needed is

```
type, public :: tree_type
   integer :: value
   type(tree_type), allocatable ::  left, right
end type tree_type
```

The `insert` Subroutine

The subroutine that inserts a new number into the tree is a straightforward implementation of the following informal recipe: if the tree is empty, make the new entry the only node of the tree; if the tree is not empty and the number to be inserted is less than the number at the root, insert the number in the left subtree; otherwise, insert the number in the right subtree.

```
recursive subroutine insert(tree, number)

   type(tree_type), allocatable, intent(in out) :: tree
   integer, intent(in) :: number
```

```
   ! If (sub)tree is empty, put number at root
   if (.not. allocated(tree)) then
      allocate (tree)
      tree%value = number
   ! Otherwise, insert into correct subtree
   else if (number < tree%value) then
      call insert(tree%left, number)
   else
      call insert(tree%right, number)
   end if

end subroutine insert
```

When a tree is allocated, the allocatable components are unallocated and there is no need to explicitly allocate them at that time.

Printing the Tree in Order

The recipe for printing the nodes of the tree follows from the way the tree has been built. It is simply to print in order the values in the left subtree of the root, print the value at the root node, then print in order the values in the right subtree. This subroutine is shown in the following complete module and program that sorts a file of integers by inserting them in an ordered binary tree, and then printing out the values in the tree in order.

```
module tree_module

   implicit none
   public :: insert, print_tree

   type, public :: tree_type
      integer :: value
      type(tree_type), allocatable ::  left, right
   end type tree_type


contains

recursive subroutine insert(tree, number)
   . . .
end subroutine insert

recursive subroutine print_tree(tree)

! Print tree in infix order

   type(tree_type), allocatable :: tree
```

```
      if (allocated(tree)) then
         call print_tree(tree % left)
         print *, tree % value
         call print_tree(tree % right)
      end if

   end subroutine print_tree

   end module tree_module

   program tree_sort

   ! Sorts a list of integers by building
   ! a tree, sorted in infix order.
   ! This sort has expected behavior n log n,
   ! but worst case (input is sorted) n ** 2.

      use tree_module
      implicit none

      ! Start with an empty tree
      type(tree_type), allocatable :: tree
      integer :: number, ios, n

      integer, dimension(*), parameter :: numbers = &
         [ 4, 6, 3, 8, 7, 9, 2, 1, 5 ]

      do n = 1, size(numbers)
         call insert(tree, numbers(n))
      end do

      call print_tree(tree)

   end program tree_sort
```

Exercises

1. Experiment with the program `tree_sort`, sorting different quantities of randomly generated integers to determine an approximate formula for the complexity of the program. It should be proportional to $n \log_2 n$.

2. Draw a tree that would be constructed by the program `tree_sort` given a list with the same numbers as above, but in the order 113, 264, 265, 265, 467, 907. What happens to the efficiency of inserting new nodes into this tree compared with the tree given in Figure 10-13?

3. Run the tree sort program in this section with two different lists, one consisting of 20,000 random numbers and the other consisting of the same numbers already sorted. Write the results to a file, rather than printing them. Compare the time taken for each run using `cpu_time`.

Input and Output 11

The facilities needed to do simple input and output tasks are described in Chapter 1, and many examples of these statements were discussed throughout the other chapters. Sometimes it is necessary to use the more sophisticated input/output features of Fortran, which are probably superior to those found in most other high-level languages. This chapter describes these features in some detail, including direct access input/output, nonadvancing input/output, unformatted input/output, the use of internal files, stream input/output, asynchronous input/output, file connection statements, the `inquire` statement, file positioning statements, and formatting.

The input/output statements are

```
read
print
write
open
close
inquire
backspace
endfile
rewind
wait
```

The `read`, `write`, and `print` statements are the ones that do the actual data transfer; the `open` and `close` statements deal with the connection between an input/output unit and a file; the `inquire` statement provides the means to find out things about a file or unit; the `backspace`, `endfile`, and `rewind` statements affect the position of the file; and the `wait` statement is used for asynchronous input/output.

Because this chapter is needed only for the more sophisticated kinds of input and output, it is organized a little bit differently from other chapters. The first part contains a discussion of some fundamental ideas needed for a thorough understanding of how Fortran input/output works. The next part of the chapter contains a description and examples of the special kinds of data transfer statements. Then there is a discussion of the `open`, `close`, `inquire`, `backspace`, `rewind`, and `endfile` statements. The final part contains a more detailed description of formatting than that provided in 1.7.

Input and output operations deal with collections of data called **files**. The data are organized into **records**, which may correspond to lines on a computer terminal, lines on a printout, or parts of a disk file. The descriptions of records and files in this chapter are to be considered abstractions and do not necessarily represent the way data are stored physically on any particular device. For example, a Fortran program may produce a file of answers. This file might be printed, and the only remaining physical representation of the file would be the ink on the paper. Or it might be written onto magnetic disk and remain there for a few years, eventually to be erased when the disk is used to store other information.

The general properties of records are discussed first.

11.1 Records

There are two kinds of records, data records and endfile records. A **data record** is a sequence of values. Thus, a data record may be represented schematically as a collection of small boxes, each containing a value, as shown in Figure 11-1.

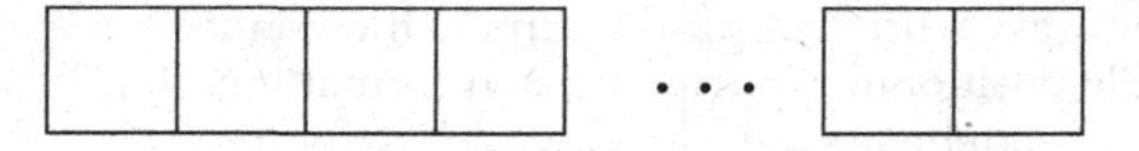

Figure 11-1 Schematic representation of the values in a record

The values in a data record may be represented in one of two ways: formatted or unformatted. If the values are characters readable by a person, each character is one value and the data are **formatted**. For example, the statement

```
write (unit=*, fmt="(i1, a, i2)") 6, ",", 11
```

would produce a record containing the four character values "6" "," "1" and "1". In this case, the record might be represented schematically as in Figure 11-2.

6	,	1	1

Figure 11-2 A formatted record with four character values

Unformatted data consist of values usually represented just as they are stored in computer memory. For example, if integers are stored using an eight-bit binary representation, execution of the statement

```
write (unit=19) 6, 11
```

might produce an unformatted record that looks like Figure 11-3.

Formatted Records

A **formatted record** is one that contains only formatted data. A formatted record may be created by a person typing at a terminal or by a program that converts values stored internally into character strings that form readable representations of those values.

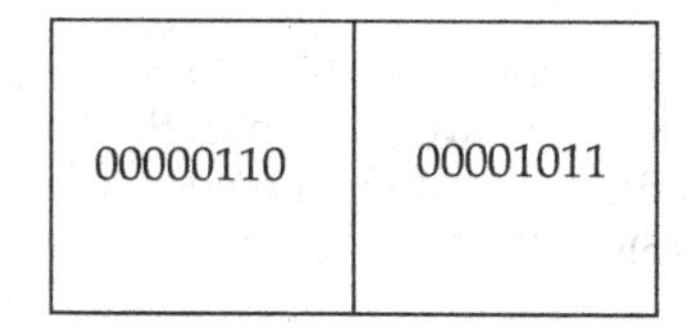

Figure 11-3 An unformatted record with two integer values

When formatted data are read into the computer, the characters must be converted to the computer's internal representation of values, which is often a binary representation. Even character values may be converted from one character representation in the record to another internal representation. The length of a formatted record is the number of characters in it; the length may be zero.

Unformatted Records

An **unformatted record** is one that contains only unformatted data. Unformatted records usually are created by running a Fortran program, although, with the knowledge of how to form the bit patterns correctly, they could be created by other means. Unformatted data often require less space on an external device. Also, it is usually faster to read and write unformatted data because no conversion is required. However, it is not as suitable for reading by humans, and usually it is not suitable for transferring data from one computer to another because the internal representation of values is machine dependent. The length of an unformatted data record depends on the number of values in it, but is measured in some processor-dependent units such as machine words; the length may be zero. The length of an unformatted record that will be produced by a particular output list may be determined by the `inquire` statement (11.6).

Endfile Records

The other kind of record is the **endfile record**, which, at least conceptually, has no values and has no length. There can be at most one endfile record in a file and it must be the last record of a file. It is used to mark the end of a file.

An endfile record may be written explicitly by the programmer using the `endfile` statement. An endfile record also is written implicitly when the last data transfer statement involving the file was an output statement, the file has not been repositioned, and

1. a `backspace` statement is executed
2. a `rewind` statement is executed or
3. the file is closed.

Record Length

In some files, the lengths of the records are fixed in advance of data being put in the file; in others, it depends on how data are written to the file. For external formatted advancing sequential output (11.3), a record ends whenever a slash (/) edit descriptor is encountered and at the conclusion of each output operation (`write` or `print`).

11.2 Files

A **file** is a collection of records. A file may be represented schematically with each box representing a record, as shown in Figure 11-4.

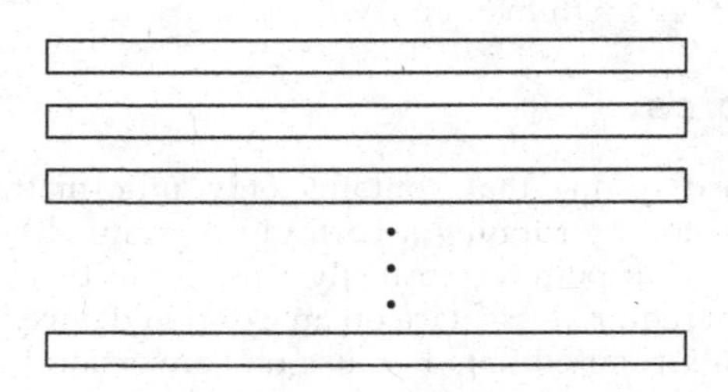

Figure 11-4 Schematic representation of records in a file

The records of a file must be either all formatted or all unformatted, except that the file may contain an endfile record as the last record. A file may have a name, but the length of the names and the characters that may be used in the names depend on the system being used. The set of names that are allowed often is determined by the operating system as well as the Fortran compiler.

A distinction is made between files that are located on an external device, such as a disk, and files in memory accessible to the program. The two kinds of files are

1. external files
2. Internal files

The use of the files is illustrated schematically in Figure 11-5.

External Files

An external file usually is stored on a peripheral device, such as a memory card, a disk, or a computer terminal. For each external file, there is a set of allowed access methods, a set of allowed forms (formatted or unformatted), a set of allowed actions, and a set of allowed record lengths. How these characteristics are established is dependent on the

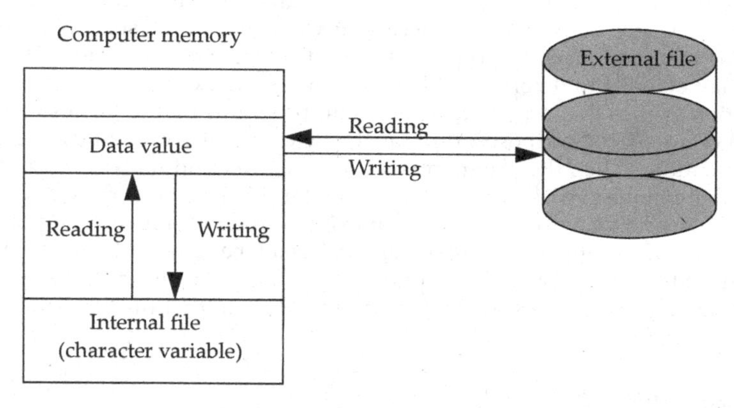

Figure 11-5 Internal and external files

computer system you are using, but usually they are determined by a combination of requests by the user of the file and actions by the operating system.

Internal Files

Internal files are stored in memory as values of character variables. The character values may be created using all the usual means of assigning character values or they may be created with an output statement using the variable as an internal file. If the variable is a scalar, the file has just one record; if the variable is an array, the file has one record for each element of the array. The length of the record is the number of characters declared or assumed for the character variable. Only formatted sequential access is permitted on internal files. For example, if `char_array` is an array of two character strings declared by

```
character(len=7), dimension(2) :: char_array
```

the statement

```
write (unit=char_array, fmt="(f7.5, /, f7.5)") 10/3.0, 10/6.0
```

produces the same effect as the assignment statements

```
char_array(1) = "3.33333"
char_array(2) = "1.66667"
```

Existence of Files

Certain files are known to the processor and are available to an executing program; these files are said to **exist** at that time. For example, a file may not exist because it is not anywhere on the disks accessible to a system. A file may not exist for a particular program because the user of the program is not authorized to access the file. Fortran programs usually are not permitted to access special system files, such as the operating system or the compiler, in order to protect them from user modification. The `inquire` statement may be used to determine whether or not a file exists.

In addition to files that are made available to programs by the processor for input, output, and other special purposes, programs may create files needed during and after program execution. When the program creates a file, it is said to exist, even if no data have been written into it. A file no longer exists after it has been deleted. Any of the input/output statements may refer to files that exist for the program at that point during execution. Some of the input/output statements (`inquire`, `open`, `close`, `write`, `print`, `rewind`, and `endfile`) can refer to files that do not exist. For example, a `write` statement can create a file that does not exist and put data into that file. An internal file always exists.

File Access Methods

There are three access methods for external files:

1. sequential access
2. direct access
3. stream access

Sequential access to the records in the file begins with the first record of the file and proceeds sequentially to the second record, and then to the next record, record by record. The records are accessed in the order that they appear in the file, as indicated in Figure 11-6. It is not possible to begin at some particular record within the file without reading from the current record down to that record in sequential order.

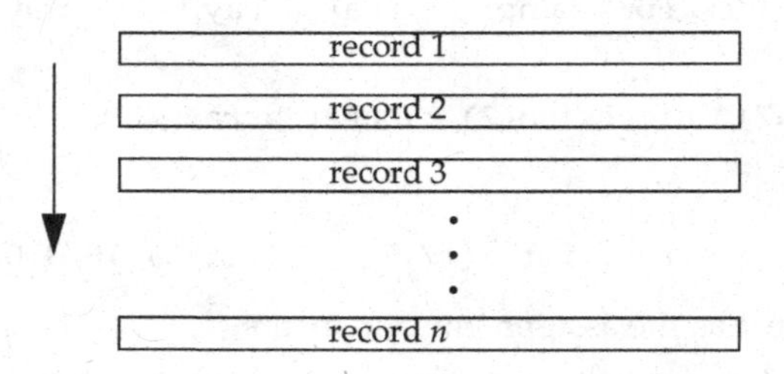

Figure 11-6 Sequential access

When a file is being accessed sequentially, the records are read and written sequentially. For example, if the records are written in any arbitrary order using direct access and then read using sequential access, the records are read beginning with record number one of the file, regardless of when it was written.

When a file is accessed directly, the records are selected by record number. Using this identification, the records may be read or written in any order. For example, it is possible to write record number 47 first, then write record number 13. In a new file, this produces a file represented by Figure 11-7. Either record may be read without first accessing the other.

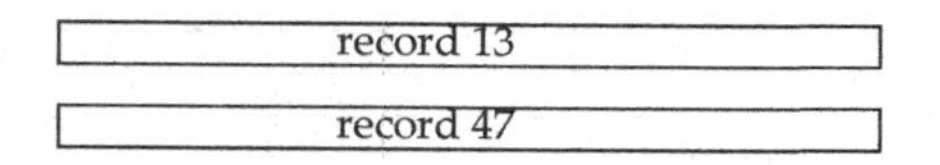

Figure 11-7 A file written using direct access

The following rules apply when accessing a file directly:

1. If a file is to be accessed directly, all the records must have the same length.
2. It is not possible to delete a record using direct access.
3. Nonadvancing input/output is prohibited.
4. An internal file must not be accessed directly.

With stream access, the file is considered to be a sequence of units, which are usually bytes. For formatted input/output, newline characters may, in effect, give a record structure to the file. There is no record structure for unformatted stream access files.

Each file has a set of permissible access methods; it is possible that a file may be accessed by more than one method. The file access method used to read or write the file must be one of the allowed access methods for the file; it is established when the file is connected to a unit (11.4). For example, the same file may be accessed sequentially by a program, then disconnected and later accessed directly by the same program, if both types of access are permitted for the file.

File Position

Each file being processed by a program has a **position**. During the course of program execution, data are read or written, causing the position of the file to change. Also, there are Fortran statements that cause the position of a file to change; an example is the `backspace` statement.

The **initial point** is the point just before the first record. The **terminal point** is the point just after the last record. These positions are illustrated by Figure 11-8. If the file is empty, the initial point and the terminal point are the same.

A file may be positioned between records. In the example pictured in Figure 11-9, the file is positioned between records 2 and 3. In this case, record 2 is the preceding record and record 3 is the next record. Of course, if a file is positioned at its initial point, there is no preceding record, and there is no next record if it is positioned at its terminal point.

There may be a current record during execution of an input/output statement or after completion of a nonadvancing input/output statement as shown in Figure 11-10, where record 2 is the current record.

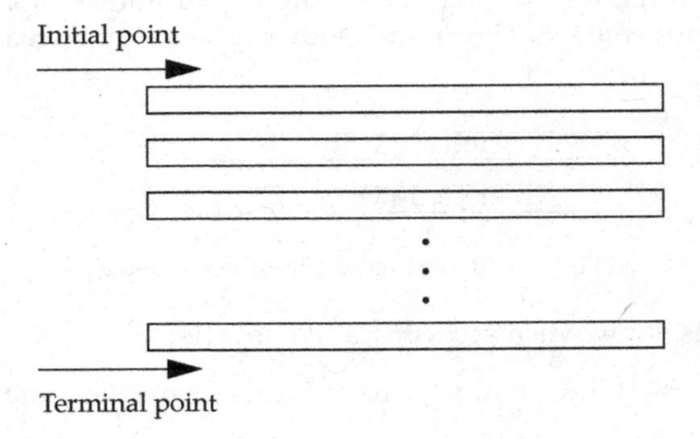

Figure 11-8 Initial and terminal points of a file

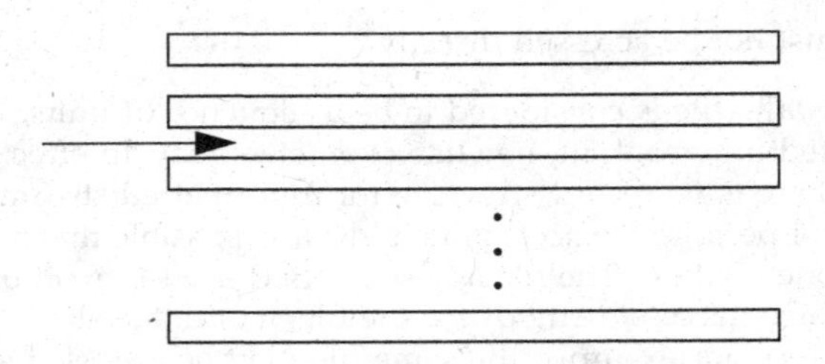

Figure 11-9 A file positioned between records

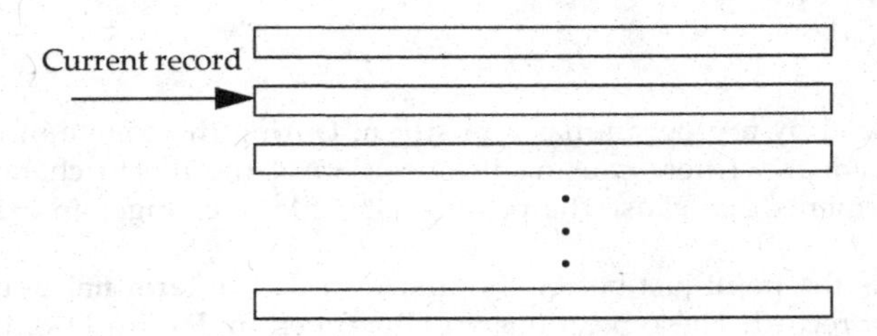

Figure 11-10 A file positioned within a current record

When there is a current record, the file is positioned at the initial point of the record, between values in a record, or at the terminal point of the record as illustrated in Figure 11-11.

An internal file is always positioned at the beginning of a record just prior to data transfer.

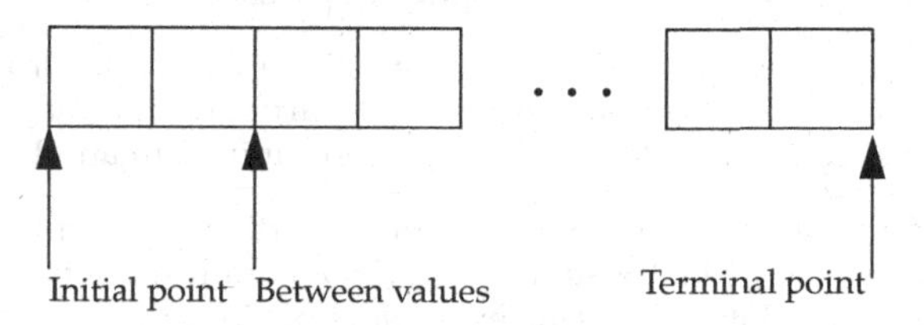

Figure 11-11 Positions within a record of a file

Advancing and Nonadvancing I/O

Advancing input/output is record oriented. Completion of an input/output operation always positions the file at the end of a record. **Nonadvancing input/output** is character oriented. After reading and writing, the file position may be between characters within a record.

Nonadvancing input/output is restricted to use with external sequential formatted files and may not be used with list-directed formatting.

Units and File Connection

Input/output statements refer to a particular file by specifying its **unit**. For example, the `read` and `write` statements do not refer to a file directly, but refer to a unit number, which must be connected to a file. The unit number for an external file is a nonnegative integer, except that it may be negative when obtained using the `newunit` specifier (11.4). The name of an internal file also is called a unit; it is a character variable. In the following examples, 5 and `char_string` are units.

```
read (unit=5) a
write (unit=char_string, fmt="(i3)") k
```

Some rules and restrictions for units are:

1. The unit * specifies a processor-determined unit number. On input, it is the same unit number that the processor would use if a `read` statement appeared without the unit number. On output, it is the same unit number that the processor would use if a `print` statement appeared without the unit number. The unit specified by an asterisk may be used only for formatted sequential access. The built-in module `iso_fortran_env` contains two integer parameters, `input_unit` and `output_unit`, which are the unit numbers used by the * unit specifier.

   Style note: Because units 5 and 6 are not universally used by Fortran compilers as the standard input and output units, use * or `input_unit` and `output_unit` instead; these last two are names of parameters in the intrinsic module `iso_fortran_env`. Note that `input_unit` and `output_unit` are allowed in the auxiliary input/output statements, such as `inquire` and `open`, but * is not.

2. File positioning, file connection, and inquiry must use external files.

3. A unit number identifies one and only one unit in a Fortran program. That is, a unit number is global to an entire program; a particular file may be connected to unit 59 in one procedure and referred to through unit 59 in another procedure.

Only certain unit numbers may be used on a particular computing system. The unit numbers that may be used are said to **exist**. Some unit numbers on some processors are always used for data input (for example, unit 5), others are always used for output (for example, unit 6). Input/output statements must refer to units that exist, except for those that close a file or inquire about a unit. The `inquire` statement may be used to determine whether or not a unit exists. On most systems units 11–99 exist and are available for general use.

To transfer data to or from an external file, the file must be connected to a unit. Once the connection is made, most input/output statements use the unit number instead of using the name of the file directly. An internal file always is connected to the unit that is the name of the character variable. There are two ways to establish connection between a unit and an external file:

1. execution of an `open` statement in the executing program

2. preconnection by the operating system

Only one file may be connected to a unit at any given time. If the unit is disconnected after its first use on a file, it may be reconnected later to another file or it may be reconnected later to the same file. A file that is not connected to a unit may not be used in any statement except the `open`, `close`, or `inquire` statements. Some units may be preconnected to files for each Fortran program by the operating system, without any action necessary by the program. For example, on many systems, units 5 and 6 are always preconnected to the default input and default output files, respectively. Preconnection of units also may be done by the operating system when requested by the user in the operating system command language. In either of these cases, the user program does not require an `open` statement to connect the file; it is preconnected.

Error, End-of-File, and End-of-Record Conditions

During execution of input/output statements, error conditions can occur. Error conditions may be checked by using the `iostat` specifier and `iomsg` specifier on many input/output statements. Each error condition will result in some positive value for the `iostat` variable, but the values used will depend on the computer system being used. Examples of errors are attempting to open a file that does not exist or attempting to read input consisting of letters when the input variable is type integer. When such an error occurs, the value of the `iostat` variable may be tested and alternative paths selected. Used with the `iostat` specifier, the `iomsg` specifier gets a character string indicating the type of problem.

If a `read` statement attempts to read an endfile record, the `iostat` variable will be set to some negative value. It will also be set to a negative value when reading beyond the end of a record with a nonadvancing **read** statement. These conditions cannot both occur at the same time.

If there is both an error condition and either an end-of-file or end-of-record condition, the `iostat` variable will be set to a positive value to indicate that an error has occurred.

The program `count_lines` counts the number of lines in a file and illustrates the use of `iostat` to determine when the end of the file is encountered. The sample run shows what happens when the input to the program is the program itself.

```
The input file is:

 program count_lines
    implicit none
    character(len=100) :: line, iom
    integer :: tally, status

    tally = 0
    print *, "The input file is:"
    print *  ! a blank line
    do
       read (unit=*, fmt="(a)", iostat=status, iomsg=iom) line
       if (status /= 0) then
          print *, trim(iom)
          exit
       end if
       write (unit=*, fmt="(t3, a)") trim(line)
       tally = tally + 1
    end do
    print *
    print *, "The file contains", tally, "lines."
 end program count_lines
End of file

The file contains 20 lines.
```

The intrinsic function `trim` removes trailing blank characters on the output lines.

11.3 Data Transfer Statements

The **data transfer** statements are the `read`, `write`, and `print` statements. In previous chapters we have seen examples of various kinds of data transfer statements. The general forms for the data transfer statements are as follows. Optional parts of a statement appear in square brackets.

`read` (*io-control-spec-list*) [*input-item-list*]
`read` *format* [, *input-item-list*]

```
write ( io-control-spec-list ) [ output-item-list ]
print format [ , output-item-list ]
```

Some examples of data transfer statements are

```
read (unit=19, iostat=is) x
write (unit=16, rec=14) y
read "(f10.2)", z
print *, zt
```

The Format Specifier

The **format specifier** (*format* in the syntax for the `print` statement and the short form of the `read` statement) may be a character expression indicating **explicit formatting** or an asterisk (*) indicating list-directed or default formatting.

The Control Information List

The input/output control specification list must contain a unit specifier of the form

```
unit= io-unit
```

and may contain at most one each of the following optional items:

```
fmt= format
rec= scalar-int-expr
iostat= scalar-default-int-variable
iomsg= scalar-char-variable
advance= scalar-char-expr
size= scalar-default-int-variable
decimal= scalar-default-char-expr
pos= scalar-int-expr
```

There are a few additional specifiers not listed here.

The input/output unit must be a nonnegative integer expression indicating an external unit connected to a file (including `input_unit` or `output_unit` from the `iso_fortran_env` module), an asterisk indicating a processor-dependent external unit, or a character variable of default kind indicating an internal unit.

The allowed forms of a format are the same within a control information list as they are in the `print` statement and the short form of the `read` statement.

There are lots of additional rules about which combinations of these items may occur; some of these rules will be covered in the discussion of various types of data transfer statements in the following sections.

The Input/Output List

The input/output list consists basically of variables in a `read` statement and expressions in a `write` or `print` statement. An implied-do loop also may appear in an input/output list.

```
print "(i9, f9.2)", (i, x(i), i = 1, n)
```

External Formatted Advancing Sequential Access I/O

The title of this section is a mouthful, but this is the kind of input/output that has been illustrated throughout the book. For formatted input and output, the file consists of characters. These characters are converted into representations suitable for storing in computer memory during input and converted from an internal representation to characters on output. When a file is accessed sequentially, records are processed in the order in which they appear in the file. Advancing input/output means that the file is positioned after the end of the last record read or written when the input/output is finished.

Templates that may be used to construct explicitly formatted sequential access data statements are

```
read ( unit= unit-number &
       , fmt= format &
       [ , iostat= scalar-default-int-variable ] &
       [ , iomsg= scalar-char-variable ] &
       [ , advance= scalar-char-expr ] &
       ) [ input-list ]
```

```
read format [ , input-list ]
```

```
write ( unit= unit-number &
        , fmt= format &
        [ , iostat= scalar-default-int-variable ] &
        [ , iomsg= scalar-char-variable ] &
        [ , advance= scalar-char-expr ] &
        ) [ output-list ]
```

```
print format [ , output-list ]
```

The *format* may be a character expression whose value is a format specification, or an asterisk indicating list-directed default formatting. For advancing input/output, the expression in the `advance` specifier must evaluate to `yes`, if it is present; nonadvancing input/output is discussed below. The `advance` specifier must not be present if the format is an asterisk designating list-directed formatting.

Examples of formatted reading are

```
read (unit=input_unit, fmt=fmt_100) a, b, c(1:40)
read (unit=19, fmt="(2f20.5)", iostat=iend, iomsg=msg) x, y
```

```
read (unit=*, fmt="(5e20.0)", advance="yes") y(1:kk)
read *, x, y
read "(2f20.5)", x, y
read *
```

Examples of formatted writing are

```
write (unit=19, fmt=fmt_103, iostat=is) a, b, c, s
write (unit=2*k, fmt=*) x
write (unit=*, fmt="(f10.5)") x
print "(a, es14.6)", " y = ", y
print *, "y = ", y
print *
```

When an advancing sequential access input/output statement is executed, reading or writing of data begins with the next character in the file. If the previous input/output statement was a nonadvancing statement, the next character transferred may be in the middle of a record, even if the statement being executed is an advancing statement. The difference between the two is that an advancing input/output statement always leaves the file positioned at the end of the record when the data transfer is completed.

The `iostat` specifier may be used to check for an end-of-file or an error condition and the `iomsg` specifier may be used to obtain an error message.

Nonadvancing Data Transfer

Like advancing input/output, the file is read or written beginning with the next character; however, nonadvancing input/output leaves the file positioned after the last character read or written, rather than skipping to the end of the record. Nonadvancing input/output is sometimes called *partial record* input/output. It may be used only with explicitly formatted external files connected for sequential access. It may not be used with list-directed input/output.

Templates that may be used to construct nonadvancing input/output statements are

```
read ( unit= unit-number &
       , fmt= format &
       , advance= scalar-char-expr &
       [ , size= scalar-default-int-variable ] &
       [ , iostat= scalar-default-int-variable ] &
       [ , iomsg= scalar-char-variable ] &
       ) [ input-list ]

write ( unit= unit-number &
       , fmt= format &
       , advance= scalar-char-expr &
       [ , iostat= scalar-default-int-variable ] &
       [ , iomsg= scalar-char-variable ] &
       ) [ output-list ]
```

The scalar character expression in the `advance` specifier must evaluate to `no` for nonadvancing input/output. The format is a character expression whose value is a format specification; it must not be an asterisk designating list-directed formatting.

The `size` variable is assigned the number of characters read on input. It does not count trailing blanks.

Examples of nonadvancing data transfer statements are

```
advancing = "no"
read (unit=15, fmt=fmt_100, advance=advancing) a, b, c
read (unit=19, fmt="(a)", advance="no",  &
      size=rec_size, iostat=ios, iomsg=message) line
write (unit=16, fmt="(i1)", advance=advancing) n
write (unit=16, fmt=fmt_200, advance="no") x(1:n)
```

The `iostat` specifier may be used to check for an end-of-file, end-of-record, or error condition and the `iomsg` specifier returns a message.

One of the important uses of nonadvancing input/output occurs when the size of the records is not known. To illustrate this, the following program counts the number of characters in a file, reading the input one character at a time. `iostat` values for end-of-record and end-of-file are required to be negative, but are otherwise processor dependent. The values –2 and –1 are typical, but the use of the intrinsic functions `is_iostat_end` and `is_iostat_eor` is preferred.

```
program char_count
   implicit none
   character(len=1) :: c
   integer :: character_count, ios
   character(len=99) :: iom

   character_count = 0
   do
      read (unit=*, fmt="(a)", advance="no", &
            iostat=ios, iomsg=iom) c
      if (ios > 0) then
         print *, trim(iom)
      else if (is_iostat_eor(ios)) then
         cycle
      else if (is_iostat_end(ios)) then
         exit
      else
         character_count = character_count + 1
      end if
   end do

   print *, "The number of characters in the file is", &
            character_count
end program char_count
```

Another obvious use of nonadvancing input/output is to print part of a line at one place in a program and finish the line later. If things are implemented properly, it also should be possible to use nonadvancing input/output to supply a prompt to a terminal and have the user type in data on the same line. This is not absolutely guaranteed, because many systems consider input from a terminal and output to the terminal to involve two different files. Here is a simple example:

```
program test_sign
   implicit none
   integer :: number
   write (unit=*, fmt="(a)", advance="no")  &
         "Type in any integer: "
   read *, number
   write (unit=*, fmt="(a, i9, a)", advance="no")  &
         "The number ", number, " is "
   if (number > 0) then
      print *, "positive."
   else if (number == 0) then
      print *, "zero."
   else
      print *, "negative."
   end if
end program test_sign
```

```
Type in any integer: 36
The number        36 is  positive.
```

Data Transfer on Internal Files

Transferring data from machine representation to characters or from characters back to machine representation can be done between two variables in an executing program. A formatted sequential access input or output statement is used; list-directed formatting is permitted. The format is used to interpret the characters. The internal file and the internal unit are the same character variable.

Templates that may be used to construct data transfer statements on an internal file are

```
read ( unit= default-char-variable &
      , fmt= format &
      [ , iostat= scalar-default-int-variable ] &
      [ , iomsg= scalar-char-variable ] &
      ) [ input-list ]
```

```
write ( unit= default-char-variable &
      , fmt= format &
      [ , iostat= scalar-default-int-variable ] &
      [ , iomsg= scalar-char-variable ] &
      ) [ output-list ]
```

Examples of data transfer statements on internal files are

```
read (unit=char_24, fmt=fmt_1, iostat=io_err) &
      mary, x, j, name
write (unit=char_var, fmt=*) x
```

Some rules and restrictions for using internal files are:

1. The unit must be a character variable that is not an array section with a vector subscript.
2. Each record of an internal file is a scalar character variable.
3. If the file is an array or an array section, each element of the array or section is a scalar character variable and thus a record. The order of the records is the order of the array elements (for arrays of rank two and greater, the first subscript varies most rapidly). The length of the record, which must be the same for each record, is the length of one array element.
4. If the character variable is an array or part of an array that has the pointer attribute, the variable must be allocated before its use as an internal file.
5. If the number of characters written is less than the length of the record, the remaining characters are set to blank. If the number of characters is greater than the length of the record, the remaining characters are truncated.
6. The records in an internal file are assigned values when the record is written. An internal file also may be assigned a value by a character assignment statement, or by some other means.
7. To read a record in an internal file, it must be defined.
8. An internal file is always positioned at the beginning before a data transfer occurs.
9. Only formatted sequential access is permitted on internal files. List-directed formatting is permitted.
10. File connection, positioning, and inquiry must not be used with internal files.
11. The use of the `iostat` and `iomsg` specifiers is the same as for external files.
12. On input, blanks are ignored in numeric fields.
13. On list-directed output, character constants are delimited with quotes.

As a simple example of the use of internal files, the following `write` statement converts the value of the integer variable `n` into the character string `s` of length 10:

```
write (unit=s, fmt="(i10)") n
```

If n = 999, the string s would be *"bbbbbbb999"*, where *b* represents a blank character. To make the conversion behave a little differently, we can force the first character of s to be a sign and make the rest of the characters digits, using as many leading zeros as necessary.

```
write (unit=s, fmt="(sp, i10.9)") n
```

Now if n = 999, the string s will have the value "+000000999".

Another use of internal input/output is to read data from a file directly into a character string, examine it to make sure it has the proper form for the data that is supposed to be read, and then read it with formatting conversion from the internal character string variable to the variables needed to hold the data. To keep the example simple, suppose that some input data record is supposed to contain ten integer values, but they have been entered into the file as ten integers separated by colons. List-directed input requires that the numbers be separated by blanks or commas. One option is to read in the data, examine the characters one at a time, and build the integers; but list-directed input will do everything except find the colon separators. So another possibility is to read in the record, change the colons to commas, and use an internal list-directed read statement to convert the character string into ten integer values. Here is a complete program, but it just reads in the numbers and prints them.

```
program p_internal

   implicit none
   character(len=100) :: internal_record
   integer, dimension(10) :: numbers
   integer :: colon_position

   read (unit=*, fmt="(a)") internal_record
   do
      colon_position = index(internal_record, ":")
      if (colon_position == 0) exit
      internal_record (colon_position:colon_position) = ","
   end do

   read (unit=internal_record, fmt=*) numbers
   print "(5i5)", numbers

end program p_internal
```

If the input is

```
3:24:53:563:-34:290:-9:883:9:224
```

the output is

```
    3   24   53  563  -34
  290   -9  883    9  224
```

Of course, in a real program, some error checking should be done to make sure that the internal record has the correct format after the colons are converted to commas.

Unformatted Input/Output

For unformatted input and output, the file usually consists of values stored using the same representation used in program memory. This means that no conversion is required during input and output. Unformatted input/output may be done using both sequential and direct access. It is always advancing.

Direct access is indicated by the presence of a `rec` specifier; sequential access occurs when no `rec` specifier is present.

Templates that may be used to construct unformatted access data statements are

```
read ( unit= unit-number &
        [ , rec= record-number ] &
        [ , iostat= scalar-default-int-variable ] &
        [ , iomsg= scalar-char-variable ] &
        ) [ input-list ]
```

```
write ( unit= unit-number &
        [ , rec= record-number ] &
        [ , iostat= scalar-default-int-variable ] &
        [ , iomsg= scalar-char-variable ] &
        ) [ output-list ]
```

Examples of unformatted access reading are

```
read (unit=18) a, b, c(1:n, 1:n)
read (unit=19, rec=14, iostat=iend) x, y
read (unit=14, iostat=k, iomsg=msg) y
```

Examples of unformatted access writing are

```
write (unit=19, iostat=ik, iomsg=im) a, b, c, s
write (unit=17, iostat=status) x
write (unit=19, rec=next_record_number) x
```

The record number given by the `rec` specifier is a scalar integer expression whose value indicates the number of the record to be read or written. If the access is sequential, the file is positioned at the beginning of the next record prior to data transfer and positioned at the end of the record when the input/output is finished, because nonadvancing unformatted input/output is not permitted. The `iostat` and `iomsg` specifiers may be used in the same way they are used for formatted input/output.

Unformatted access is very useful when creating a file of data that must be saved from one execution of a program and used for a later execution of the program. Suppose, for example that a program deals with the inventory of a large number of automobile parts. The data for each part (in a module in our simple example) consists of the part number and the quantity in stock.

```
type, public :: part
   integer :: id_number, qty_in_stock
end type part

type (part), dimension(10000), public :: part_list
integer, public:: number_of_parts
```

Suppose the integer variable number_of_parts records the number of different parts that are stored in the array part_list. At the end of the program, the number of parts and the entire part list can be saved in the file named part_file with the following statements:

```
open (unit=19, file="part_file", position="rewind", &
      form="unformatted", action="write", status="replace")
write (unit=19), number_of_parts, part_list(1:number_of_parts)
```

At the beginning of the next execution of the program, the inventory can be read back into memory with the statements:

```
open (unit=19, file="part_file", position="rewind", &
      form="unformatted", action="read", status="old")
read (unit=19), number_of_parts, part_list(1:number_of_parts)
```

See 11.4 for the description of the open statement.

Direct Access Data Transfer

When a file is accessed directly, the record to be processed is the one given by the record number in a rec specifier. The file may be formatted or unformatted.

Templates that may be used to construct direct access data statements are

```
read ( unit= unit-number &
       , [ fmt= format ] &
       , rec= record-number &
       [ , iostat= scalar-default-int-variable ] &
       [ , iomsg= scalar-char-variable ] &
       ) [ input-list ]

write ( unit= unit-number ] &
       [ , fmt= format ] &
       , rec= record-number &
       [ , iostat= scalar-default-int-variable ] &
       [ , iomsg= scalar-char-variable ] &
       ) [ output-list ]
```

The *format* must not be an asterisk.

Examples of direct access input/output statements are

```
read (unit=input_unit, fmt=fmt_x, rec=32) a
read (unit=12, rec=34, iostat=io_status) a, b, d
write (unit=output_unit, fmt="(2f15.5)", rec=n+2) x, y
```

The iostat and iomsg specifiers are used just as they are with sequential access.

To illustrate the use of direct access files, let us consider the simple automobile parts example used to illustrate unformatted input/output. In this example, suppose that the parts list is so large that it is not feasible to read the entire list into memory. Instead, each time information about a part is needed, just the information about that one part is read from an external file. To do this in a reasonable amount of time, the file must be stored on a device such as a disk, where each part is accessible as readily as any other. Analogous but more realistic examples might involve the bank accounts for all customers of a bank or tax information on all tax-payers in one country. This time a structure is not needed, because the only information in the file is the quantity on hand. The part identification number is used as the record number of the record in the file used to store the information for the part having that number. Also, the array is not needed because the program deals with only one part at a time.

Suppose we just need a program that looks up the quantity in stock for a given part number. This program queries the user for the part number, looks up the quantity on hand by reading one record from a file, and prints out that quantity.

```
program part_info
   implicit none
   integer :: part_number, qty_in_stock

   print *, "Enter part number"
   read *, part_number
   open (unit=19, file="part_file", access="direct", recl=10,  &
         form="unformatted", action="read", status="old")
   read (unit=19, rec=part_number) qty_in_stock
   print *, "The quantity in stock is", qty_in_stock
end program part_info
```

Of course, the program could be a little more sophisticated by using a loop to repeat the process of asking for a part number and providing the quantity in stock. Also, there must be other programs that create and maintain the file that holds the database of information about the parts. A more complex organization for the file may be necessary if the range of legal part numbers greatly exceeds the actual number of different parts for which information is saved.

Stream Access Data Transfer

When data transfer occurs with a file opened for stream access, the data transfer statements are the same as for either sequential or direct access, except that the file may not use nonadvancing input/output. Additionally, it may use the pos specifier to indicate a position in the stream file at which to begin the data transfer. The value of the pos specifier must be one that was returned by the pos specifier of an inquire statement. The following example shows how stream access and the pos specifier work.

```
program stream

   implicit none
   integer :: ios, mark_pos
   real :: x
   character(len=99) :: iom = "OK"

   open (unit=11, file="saf", access="stream", &
         form="unformatted", iostat=ios, iomsg=iom)
   print *, iom
   write (unit=11) 1.1
   write (unit=11) 2.2

   inquire (unit=11, pos=mark_pos)
   write (unit=11) 3.3
   write (unit=11) 4.4

   read (unit=11, pos=mark_pos) x
   print *, x

end program stream
```

The open statement indicates stream access and when it is executed the value of iom is not changed from its initial value if there is no error. The first two write statements put 1.1 and 2.2 in the file and then the inquire statement records the position of the file as just after the 2.2. After two more write statements, x is read from the position recorded by mark_pos resulting in its value being 3.3. The output from the program is

```
OK
3.3
```

Asynchronous Input/Output

Normally, execution waits for an input/output statement to complete. Asynchronous input/output allows computation to proceed while the input/output is taking place. The **wait statement** allows for the program to wait at a certain point until the input/output is complete. The inquire statement (11.6) may be used to determine if a particular file may be read or written asynchronously and then the open statement (11.4) may specify that asynchronous input/output is going to be used.

In this simple example, the array data is rank 3. The third subscript ranges from 1 to 13 and we want to read the values for a single subscript and process that data while data is being read for the next value of the subscript.

The lines below suggest how the program executes. The values for subscript 1 are read, then the program waits for that to finish and starts reading the values for subscript 2. While that input is taking place the values for subscript 1 are processed. Then the program waits for the input of values for subscript 2 to complete. And so forth.

```
read 1
======
wait 1
read 2
process 1
======
wait 2
read 3
process 2
======
 . . .
======
wait 12
read 13
process 12
======
wait 13
process 13
```

```
inquire (file="data_file", asynchronous=async_ok)
if (async_ok == "YES") then
   open (unit=11, file="data_file", &
         asynchronous="yes", iostat=ios, &
         form="unformatted", &
         position="rewind", action="readwrite")
   . . .
read (unit=11, asynchronous="yes") data(:,:,1)
do i = 1, 12
   wait (unit=11)
   read (unit=11, asynchronous="yes") data(:,:,i+1)
   call process(data(:,:,i))
end do
wait (unit=11)
call process(data(:,:,13))
```

11.4 The open Statement

The **open statement** establishes a connection between a unit and an external file and determines the connection properties. After this is done, the file can be used for data transfers (reading and writing) using the unit number. It is not necessary to execute an **open** statement for files that are preconnected to a unit.

The open statement may appear anywhere in a program and, once executed, the connection of the unit to the file is valid in the main program or any subprogram for the remainder of that execution, unless a close statement affecting the connection is executed.

If a file is connected to one unit, it may not be connected to a different unit at the same time.

Execution of an open statement using a unit that is already open is permitted in only a few special circumstances.

Syntax Rules for the open Statement

The form of the open statement is

open (*connect-spec-list*)

where some of the permissible connection specifications are

unit= *external-file-unit*
newunit= *scalar-int-variable*
file= *file-name-expr*
access= *scalar-char-expr*
action= *scalar-char-expr*
asynchronous= *logical*
decimal= *scalar-char-expr*
encoding= *scalar-char-expr*
form= *scalar-char-expr*
iostat= *scalar-int-variable*
iomsg= *scalar-char-variable*
position= *scalar-char-expr*
recl= *scalar-int-expr*
status= *scalar-char-expr*

Examples are

```
open (unit=19, iostat=ios, iomsg=iom, &
      status="scratch", action="readwrite")
open (unit=18, access="direct", file="plot_data", &
      status="old", action="read")

open (newunit=unit_number, file="xxx", status="replace")
write (unit=unit_number, fmt= . . .
```

Some rules, restrictions, and suggestions for the open statement are:

1. An external unit number or the newunit specifier is required.
2. A specifier may appear at most once in any open statement.

3. The `file` specifier must appear if the `status` is `old`, `new`, or `replace`; the `file` specifier must not appear if the `status` is `scratch`.
4. The `status` specifier should appear.
5. The `action` specifier should appear. It must not be `read` if the status is `new` or `replace`. It should be `readwrite` if the status is `scratch`.
6. The `position` specifier should appear if the access is `sequential` and the status is `old`.
7. The character expression established for many of the specifiers must contain permitted values in a list of alternative values as described below. For example, `old`, `new`, `replace`, and `scratch` are permitted for the `status` specifier. Uppercase letters may be used. Trailing blanks in any specifier are ignored.

The Connection Specifiers

`access` The value of the `access` specifier must be `direct`, `sequential`, or `stream`. The method must be an allowed access method for the file. If the file is new, the allowed access methods given to the file must include the one indicated. If the access is direct, there must be a `recl` specifier to specify the record length.

`action` The value of the `action` specifier must be `read`, `write`, or `readwrite`. `read` indicates that `write`, `print`, and `endfile` statements are prohibited. `write` indicates that `read` statements are prohibited. `readwrite` indicates that any input/output statement is permitted.

`asynchronous` The value of the `asynchronous` specifier must be `yes` or `no`. It indicates whether asynchronous input/output may be used with the file.

`decimal` The value must be `comma` or `point`, indicating the decimal symbol.

`encoding` The value must be `utf-8` or `default`, which indicates the encoding method for formatted files.

`file` The `file` specifier indicates the name of the file to be connected. If the name is omitted, the connection will be made to a processor-determined file.

`form` The value of the `form` specifier must be `formatted` or `unformatted`. `formatted` indicates that all records will be formatted. `unformatted` indicates that all records will be unformatted. If the file is connected for direct access or stream access, the default is `unformatted`. If the file is connected for sequential access, the default is `formatted`. If the file is new, the allowed forms given to the file must include the one indicated.

`iomsg` The value returned is an explanatory message if an error occurs. It has no effect unless there is an `iostat` specifier.

`iostat` The `iostat` specifier must be an integer variable. It is given a value that is a positive integer if there is an error condition while executing the `open` statement and zero if there is no error.

`newunit` The `newunit` specifier must be an integer variable. It is set to a value less than –1 that is not already used as a unit number. This number then may be used as a unit number, knowing it will not conflict with any other unit number in the program.

`position` The value of the `position` specifier must be `rewind` or `append`. `rewind` positions the file at its initial point. `append` positions the file at the terminal point or just before an endfile record, if there is one. The file must be connected for sequential access. If the file is new, it is positioned at its initial point.

`recl` The `recl` specifier has a positive value that specifies the length of each record if the access method is `direct` or the maximum length of a record if the access method is `sequential`. If the file is connected for formatted input/output, the length is the number of characters. If the file is connected for unformatted input/output, the length is measured in processor-dependent units. The length may be, for example, the number of computer words. If the file exists, the length of the record specified must be an allowed record length. If the file does not exist, the file is created with the specified length as an allowed length.

`status` The value of the `status` specifier should be `old`, `new`, `replace`, or `scratch`. `old` refers to a file that must exist. `new` refers to a file that must not exist. If the status is `replace` and the file does not exist, it is created and given a status of `old`. If the status is `replace` and the file does exist, it is deleted, a new file is created with the same name, and its status is changed to `old`. `scratch` refers to a scratch file that exists only until termination of execution of the program or until a `close` is executed on that unit. Scratch files must be unnamed. `replace` is recommended when it is not certain if there is an old version, but it is to be replaced if there is one.

Exercise

1. Open a file using a new unit. Print the value of the unit number. Open a second file using a new unit. Print the value of the unit number. Close the second file. Open a third file using a new unit. Print the value of the unit number.

11.5 The `close` Statement

Execution of a **`close` statement** terminates the connection of a file to a unit. Any connections not closed explicitly by a `close` statement are closed by the operating system when the program terminates. The form of the `close` statement is

`close (` *close-spec-list* `)`

The items in the close specification list may be selected from

`unit=` *external-file-unit*
`iostat=` *scalar-default-int-variable*
`iomsg=` *scalar-default-char-variable*
`status=` *scalar-char-expr*

Examples are

```
close (unit=19)
close (unit=18, iostat=ir, iomsg=im, status="keep")
```

Some rules for the `close` statement are

1. An external unit number (including possibly `input_unit` or `output_unit`) is required.
2. A `close` statement may refer to a unit that is not connected or does not exist, but it has no effect. This is not considered an error.
3. The `status` specifier must have a value that is `keep` or `delete`. If it is `keep`, the file continues to exist after closing the file. If it has the value of `delete`, the file will not exist after closing the file. The default value is `keep`, except that the default for `scratch` files is `delete`. If you specify a `status` for a `scratch` file, it must be `delete`.
4. The rules for the `iostat` and `iomsg` specifiers are the same as for the `open` statement.
5. A specifier must not appear more than once in a `close` statement.
6. Connections that have been closed may be reopened at a later point in an executing program. The new connection may be to the same or to a different file.

11.6 The `inquire` Statement

The **`inquire` statement** provides the capability of determining information about a unit's or file's existence, connection, access method, or other properties during execution of a program. For each property inquired about, a scalar variable of default kind is supplied; that variable is given a value that answers the inquiry. The variable may be tested and optional execution paths selected in a program based on the answer returned. The inquiry specifiers are indicated by keywords in the `inquire` statement. A file inquiry may be made by unit number, by the file name, or by an output list that might be used in an unformatted direct access output statement.

The values of the character items (except `name`) are always in uppercase.

Syntax Rules for the `inquire` Statement

The form of an inquiry by unit number or file name is

`inquire (` *inquiry-spec-list* `)`

An **inquiry by unit** must include the following in the inquiry specification list:

`unit=` *external-file-unit*

An **inquiry by name** must include the following in the inquiry specification list:

`file=` *file-name*

The expression for the file name may refer to a file that is not connected or does not exist. The value for the file name must be acceptable to the system. An `inquire` statement must not have both a file specifier and a unit specifier.

In addition, the inquiry specification list may contain the following items and a few more not listed here. The type of the item following the keyword is indicated; each item following the keyword and equals sign must be a scalar variable of default kind, if it is not type integer.

`access=` *character*
`action=` *character*
`asynchronous=` *logical*
`decimal=` *character*
`direct=` *character*
`encoding=` *character*
`exist=` *logical*
`form=` *character*
`formatted=` *character*
`iomsg=` *character*
`iostat=` *integer*
`name=` *character*
`named=` *logical*
`nextrec=` *integer*
`number=` *integer*
`opened=` *logical*
`pos=` *character*

```
position= character
read= character
readwrite= character
recl= integer
sequential= character
stream= character
unformatted= character
write= character
```

Examples of the `inquire` statement are

```
inquire (unit=19, exist=ex)
inquire (file="t123", opened=op, access=ac)
```

The `iolength` *Inquiry*

The form of an `inquire` statement used to determine the length of an output item list is

```
inquire ( iolength= scalar-default-int-variable ) output-item-list
```

The length value returned in the scalar integer variable will be an acceptable value that can be used later as the value of the `recl` specifier in an `open` statement to connect a file whose records will hold the data indicated by the output list of the `inquire` statement.

An example of this form of the `inquire` statement is

```
inquire (iolength=iolen) x, y, cat
```

Specifiers for Inquiry by Unit or File Name

This section describes the syntax and effect of the inquiry specifiers that may appear in the unit and file forms of the `inquire` statement. The values returned in the inquiry specification list are those current at that point in the execution of the program.

The `iostat` inquiry specifier indicates error condition information about the inquiry statement execution itself. If an error condition occurs, all the inquiry specifiers are undefined except the `iostat` and `iomsg` specifiers.

`access` The value returned is `SEQUENTIAL` if the file is connected for sequential access, `DIRECT` if the file is connected for direct access, `STREAM` if the file is connected for stream access, or `UNDEFINED` if the file is not connected.

`action` The value returned is `READ` if the file is connected limiting the access to input, `WRITE` if the file is connected limiting the access to output, `READWRITE` if the file is connected for input and output, or `UNDEFINED` if the file is not connected.

`asynchronous`	The character variable indicates whether the connection allows asynchronous input/output.
`decimal`	The value returned is `COMMA` if the decimal symbol is a comma, `POINT` if the decimal symbol is a period, and `UNDEFINED` if the unit or file is not connected for formatted input/output.
`direct`	The value returned is `YES` if direct access is an allowed method, `NO` if direct access is not an allowed method, or `UNKNOWN` if the processor does not know if direct access is allowed.
`encoding`	The character encoding used for the file.
`exist`	If the inquiry is by unit, the logical variable indicates whether or not the unit exists. If the inquiry is by file, the logical variable indicates whether or not the file exists.
`form`	The value returned is `FORMATTED` if the file is connected for formatted input/output, `UNFORMATTED` if the file is connected for unformatted input/output, or `UNDEFINED` if the file is not connected.
`formatted`	The value returned is `YES` if formatted input/output is permitted for the file, `NO` if formatted input/output is not permitted for the file, or `UNKNOWN` if the processor cannot determine if formatted input/output is permitted for the file.
`iolength`	The record length for direct-access input/output. This is used only in an inquire by length inquiry.
`iomsg`	The value returned is an explanatory message if an error occurs. It has no effect if `iostat` is not present.
`iostat`	The `iostat` specifier must be an integer variable. It is given a value that is a positive integer if there is an error condition while executing the `open` statement and zero if there is no error.
`name`	The value is the name of the file if the file has a name; otherwise, the designated variable becomes undefined. The processor may return a name different from the one given in the `file` specifier by the program, because a user identifier or some other processor requirement for file names may be added. The name returned will be acceptable for use in a subsequent `open` statement. The case (upper or lower) used is determined by the processor.
`named`	The scalar logical value is true if and only if the file has a name.
`nextrec`	The integer value returned is one more than the last record number read or written in a file connected for direct access. If no records have been processed, the value is 1. The specified variable becomes unde-

fined if the file is not connected for direct access or if the file position is indeterminate because of a previous error condition.

`number` The value returned is the number of the unit connected to the file. If there is no unit connected to the file, the value is –1.

`opened` If the inquiry is by unit, the logical variable indicates whether or not the unit is connected to some file. If the inquiry is by file, the logical variable indicates whether or not the file is connected to some unit.

`pos` The value returned is the position (usually byte number) of a file connected for stream access, and is undefined otherwise.

`position` The value returned is REWIND if the file is connected with its position at the initial point, APPEND if the file is connected with its position at the end point or UNDEFINED if the file is not connected, is connected for direct access or if any repositioning has occurred since the file was connected.

`read` The value returned is YES if read is one of the allowed actions for the file, NO if read is not one of the allowed actions for the file, or UNKNOWN if the processor is unable to determine if read is one of the allowed actions for the file.

`readwrite` The value returned is YES if input and output are allowed for the file, NO if input and output are not both allowed for the file, or UNKNOWN if the processor is unable to determine if input and output are allowed for the file.

`recl` The integer value returned is the maximum record length of the file. For a formatted file, the length is in characters. For an unformatted file, the length is in processor-dependent units. If the file does not exist, the specified variable becomes undefined.

`sequential` The value returned is YES if sequential access is an allowed method, NO if sequential access is not an allowed method, or UNKNOWN if the processor does not know if sequential access is allowed.

`stream` The value returned is YES if stream access is an allowed method, NO if stream access is not an allowed method, or UNKNOWN if the processor does not know if stream access is allowed.

`unformatted` The value returned is YES if unformatted input/output is permitted for the file, NO if unformatted input/output is not permitted for the file, or UNKNOWN if the processor cannot determine if unformatted input/output is permitted for the file.

`write` The value returned is YES if write is one of the allowed actions for the file, NO if write is not one of the allowed actions for the file, or UNKNOWN if the processor is unable to determine if `write` is one of the allowed actions for the file.

Table of Values Assigned by `inquire`

Table 11-1 indicates the values assigned to the various variables by the execution of an `inquire` statement.

Table 11-1 Values assigned by the inquire statement

	Inquire by file		Inquire by unit	
Specifier	**Unconnected**	**Connected**	**Connected**	**Unconnected**
`access`	UNDEFINED	SEQUENTIAL or DIRECT		UNDEFINED
`action`	UNDEFINED	READ, WRITE, or READWRITE		UNDEFINED
`asynchronous`	UNDEFINED	YES or NO		UNDEFINED
`decimal`	UNDEFINED	COMMA, POINT, or UNDEFINED		UNDEFINED
`direct`	UNKNOWN	YES, NO, or UNKNOWN		UNKNOWN
`encoding`	UTF-8, UNKNOWN,or other	UTF-8, UNDEFINED, or UNKNOWN		UNKNOWN
`exist`	True if file exists, false otherwise		True if unit exists, false otherwise	
`form`	UNDEFINED	FORMATTED or UNFORMATTED		UNDEFINED
`formatted`	UNKNOWN	YES, NO, or UNKNOWN		UNKNOWN
`iolength`	*recl* value for *output-item-list*			
`iomsg`	Error message or unchanged			Unchanged
`iostat`	0 for no error, a positive integer for an error			
`name`	Filename (may not be same as file=value)		Filename if named, else undefined	Undefined
`named`	True		True if file named, false otherwise	False
`nextrec`	Undefined	If direct access, next record #; else undefined		Undefined
`number`	−1	Unit number		−1
`opened`	False	True		False

Table 11-1 (Continued) Values assigned by the inquire statement

pos	Undefined value	Stream file position or undefined value	Undefined value
position	UNDEFINED	REWIND, APPEND, ASIS, or UNDEFINED	UNDEFINED
read	UNKNOWN	YES, NO, or UNKNOWN	UNKNOWN
readwrite	UNKNOWN	YES, NO, or UNKNOWN	UNKNOWN
recl	Undefined	If direct access, record length; else maximum record length	Undefined
sequential	UNKNOWN	YES, NO, or UNKNOWN	UNKNOWN
stream	UNKNOWN	YES, NO, or UNKNOWN	UNKNOWN
unformatted	UNKNOWN	YES, NO, or UNKNOWN	UNKNOWN
write	UNKNOWN	YES, NO, or UNKNOWN	UNKNOWN

11.7 File Positioning Statements

Execution of a data transfer usually changes the position of a file. In addition, there are three statements whose main purpose is to change the position of a file. Changing the position backwards by one record is called **backspacing**. Changing the position to the beginning of the file is called **rewinding**. The `endfile` statement writes an endfile record and positions the file after the endfile record.

The syntax of the file positioning statements is

```
backspace ( position-spec-list )
rewind ( position-spec-list )
endfile ( position-spec-list )
```

A position specification may be either of the following:

```
unit= external-file-unit
iostat= scalar-default-int-variable
iomsg= scalar-default-char-variable
```

Examples of file positioning statements are

```
backspace (unit=18, iostat=status)
rewind (unit=input_unit)
endfile (unit=11, iostat=ierr, iomsg=endfile_error_message)
```

Rules and restrictions for file positioning statements:

1. The `backspace`, `rewind`, and `endfile` statements may be used only to position external files.
2. The external file unit number is required.
3. The files must be connected for sequential access.

The *backspace* Statement

Execution of a **backspace statement** causes the file to be positioned before the current record if there is a current record, or before the preceding record if there is no current record. If there is no current record and no preceding record, the position of the file is not changed. If the preceding record is an endfile record, the file becomes positioned before the endfile record. If a `backspace` statement causes the implicit writing of an endfile record and if there is a preceding record, the file becomes positioned before the record that precedes the endfile record.

If the file is already at its initial point, a `backspace` statement does not affect it. Backspacing over records written using list-directed formatting is prohibited.

The *rewind* Statement

A **rewind statement** positions the file at its initial point. Rewinding has no effect on the position of a file already at its initial point.

The *endfile* Statement

The **endfile statement** writes an endfile record and positions the file after the endfile record. Writing records past the endfile record is prohibited. After executing an `endfile` statement, it is necessary to execute a `backspace` or a `rewind` statement to position the file before the endfile record prior to reading or writing the file.

11.8 Formatting

Data usually are stored in memory as the values of variables in some binary form. For example, the integer 6 may be stored as 0000000000000110, where the 1s and 0s represent bits. On the other hand, formatted data records in a file consist of characters. Thus, when data are read from a formatted record, it must be converted from characters to the internal representation, and when data are written to a formatted record, it must be converted from the internal representation into a string of characters. A **format specification** provides the information needed to determine how these conversions are to be performed. The format specification is basically a list of **edit descriptors** that describe the format for the values in the input/output list.

A format specification is written as a character string or more complicated character expression. The expression, when evaluated, must be a valid format specification. Using these methods is called **explicit formatting.**

There is also list-directed formatting. Formatting (that is, conversion) occurs without specifically providing the editing information usually contained in a format specification. In this case, the editing or formatting is implicit. List-directed editing, also called "default formatting", is explained below.

Some rules and restrictions pertaining to format specifications are:

1. If the expression is a character array, the format is derived in array element order.
2. If the expression is an array element, the format must be entirely contained within that element.

Format Specifications

The items that make up a format specification are **edit descriptors**, which may be **data edit descriptors** or **control edit descriptors**. Each data list item must have a corresponding data edit descriptor; other descriptors control spacing, tabulation, etc.

Each format item has one of the following forms:

[*r*] *data-edit-desc*
control-edit-desc
[*r*] (*format-item-list*)
* (*format-item-list*)

where *r* is an integer literal constant called a **repeat factor**; it must be a positive integer with no kind parameter value. The asterisk (*) indicates an unlimited format item list, which behaves essentially as a very large number.

Examples:

```
read (unit=*, fmt="(5e10.1, i10)") max_values, k
print "(a, 2i5)", "The two values are: ", n(1), n(2)
```

The data edit descriptors have the forms shown in Table 11-2, where *w* specifies the width of the field, *m* the minimum number of digits written, *d* the number of decimal places, and *e* the number of digits in the exponent. There are other edit descriptors that have functionality that duplicates those given in the table.

Table 11-2 Data edit descriptors

Descriptor	Data type
`i` *w* [. *m*]	Decimal integer
`f` *w* . *d*	Real, positional form
`es` *w* . *d* [e *e*]	Real, scientific form
`en` *w* . *d* [e *e*]	Real, engineering form

Table 11-2 *(Continued)* Data edit descriptors

Descriptor	Data type
`l` *w*	Logical
`a` [*w*]	Character ("alphabetic")
`dt` [*type* [`(` *args* `)`]]	Derived type

w, m, d, and *e* must be integer literal constants with no kind parameter. *w* must be nonnegative for the `i` and `f` edit descriptors; it must be positive for the `es`, `l`, and `a` edit descriptors. *e* must be positive. *d* and *m* must be nonnegative. The values of *m, d,* and *e* must not exceed the value of *w.*

When *w* is not zero, it designates the width of the field in the file, that is, the number of characters transferred to or from the file. When *w* is zero, the value is printed in the smallest field in which it fits. Also, as explained in the description of each edit descriptor, *m* is the number of digits in the number field, *d* is the number of digits after the decimal point, and *e* is the number of digits in the exponent.

For derived-type editing, the optional *type* and *args* are passed to a subroutine performing the data transfer.

The control edit descriptors have the forms shown in Table 11-3. *n* is a positive integer literal constant with no kind parameter.

Table 11-3 Control Edit Descriptors

Descriptor	Function
`t` *n*	Tab to position *n*
`tl` *n*	Tab left *n* positions
`tr` *n*	Tab right *n* positions
[*n*] `/`	Next record
`:`	Stop formatting if i/o list is exhausted
`s`	Default printing of plus sign
`sp`	Print plus sign
`ss`	Suppress plus sign
`dc`	Decimal point symbol is a comma
`dp`	Decimal point symbol is a period

Formatted Data Transfer

When formatted data transfer is taking place, the next item in the input/output list is matched with the next data edit descriptor to determine the form of conversion between the internal representation of the data and the string of characters in the formatted record. Before this matching process occurs, the input/output list is considered to be expanded by writing out each element in an array and each component in a structure, unless a dt edit descriptor matches the structure. Analogously, the repeated edit descriptors are considered to be expanded, and the whole specification is considered to be repeated as often as necessary to accommodate the entire list, as explained below regarding the use of parentheses. Let us take an example:

```
print "(i5, 2(i3, tr1, i4), i5)", i, n(1:4), j
```

The expanded input/output list would be

```
i, n(1), n(2), n(3), n(4), j
```

and the expanded list of edit descriptors would be

```
i5, i3, tr1, i4, i3, tr1, i4, i5
```

As the formatting proceeds, each input/output list item is read or written with a conversion as specified by its corresponding data edit descriptor. Note that complex data type items require two real data edit descriptors. The control edit descriptors affect the editing process at the point they occur in the list of edit descriptors.

An empty format specification such as ( ) is restricted to input/output statements without a list of items. The effect is that no characters are read or written.

Control edit descriptors do not require a corresponding data item in the list. When the data items are completely processed, any control edit descriptors occurring next in the expanded list of edit descriptors are processed and then the formatting terminates.

Parentheses in a Format Specification

The action indicated by encountering a right parenthesis in a format specification depends on the nesting level. The rules are:

1. When the rightmost right parenthesis is encountered and there are no more data items, input/output terminates.

2. When the rightmost right parenthesis is encountered and there are more data items, the format is searched backward first until a right parenthesis is encountered, then back to the matching left parenthesis. If there is no other right parenthesis except the outermost one, format control reverts to the left parenthesis at the beginning of the format specification. A slash edit descriptor is considered to occur after processing the rightmost right parenthesis and before processing the left parenthesis.

3. If there is a repeat factor or asterisk encountered to the left of the left parenthesis, the repeat before the parenthesis is reused.

4. Sign control is not affected. It remains in effect until another sign edit descriptor is encountered.

This process is illustrated by the following two cases:

```
print "(a, i5)", "x", 1, "y", 2, "z", 3
```

is equivalent to

```
print "(a, i5, /, a, i5, /, a, i5)", "x", 1, "y", 2, "z", 3
```

and

```
print "(a, (i5))", "x", 1, 2, 3
```

is equivalent to

```
print "(a, i5, /, i5, /, i5)", "x", 1, 2, 3
```

An unlimited format list and the colon (`:`) edit descriptor provide a way to print a comma-separated list; the colon prevents printing a comma after the last number.

```
program unlimited_format

   implicit none

   integer, dimension(:), allocatable :: A
   integer :: n, i

   print *, "Enter a number"
   read *, n
   A = [ (i, i = 1, n) ]

   print "(a, *(i0, :, ",", tr1))", "A = ", A

end program unlimited_format
```

```
 Enter a number
6
A = 1, 2, 3, 4, 5, 6
```

Numeric Editing

The edit descriptors that cover numeric editing are `i`, `f`, `en`, and `es`. The following rules apply to all of them.

On input:

1. Leading blanks are not significant.
2. Within a field, blanks are ignored.
3. Plus signs may be omitted in the input data.
4. In numeric fields that have a decimal point and correspond to `f`, `en`, or `es` edit descriptors, the decimal point in the input field overrides placement of the decimal point indicated by the edit descriptor.

5. There must be no kind parameter in the input data.

On output:

1. A positive or zero internal value may have a plus sign, depending on the sign edit descriptors used.
2. The number is right justified in the field. Leading blanks may be inserted.
3. If the number or the exponent is too large for the field width specified in the edit descriptor, the output field is filled with asterisks. The processor must not produce asterisks when elimination of optional characters (such as the optional plus sign indicated by the `sp` edit descriptor) will allow the output to fit into the output field.

Integer Editing

The integer edit descriptor is

`i` *w* [. *m*]

w is the field width unless *w* = 0, in which case the minimal necessary field width is used; *m* is the least number of digits to be output. *m* has no effect on an input field. If *m* is omitted, its default value is 1. The value of *m* must not exceed the value of *w* unless *w* = 0. Leading zeros pad an integer field to the value of *m*. The field on output consists of an optional sign and the magnitude of the integer number without leading zeros, except in the case of padding to the value of *m*. All blanks are output only if the magnitude is zero and *m* = 0.

Input: The character string in the file must be an optionally signed integer constant.

Output: The field consists of leading blanks, followed by an optional sign, followed by the unsigned value of the integer. At least one digit must be output unless *m* is 0 and the output value is 0.

Example:

```
read (unit=15, fmt="(i5, i8)") i, j
```

If the input field is

*bbb*2401110107

`i` is read using the integer `i5` edit descriptor and `j` is read with a `i8` edit descriptor. The resulting values of `i` and `j` are 24 and 1,110,107, respectively.

Example:

```
integer :: i
integer, dimension(4), parameter :: ik = [2, 1, 0, 0]
character(len=*), parameter :: &
      i_format = "(a, i3, a, i3.3, a, i0, a, i0.0, a)"
print i_format, "|", (ik(i), "|", i=1,4)
```

produces the line of output:

```
| 2|001|0| |
```

Real Editing

The forms of the edit descriptors for real values are

```
fw.d
esw.d [ee ]
enw.d [ee ]
```

The f Edit Descriptor

If $w > 0$, `f` editing converts to or from a string of w digits with d digits after the decimal point. d must not be greater than w. The number may be signed. If $w = 0$, the value is printed using the smallest field width possible; w may not be 0 for input.

Input: If the input field contains a decimal point, the value of d has no effect. If there is no decimal point, a decimal point is inserted in front of the rightmost d digits. There may be more digits in the number than the processor can use. On input, the number may contain an `e` indicating an exponent value.

Output: The number is an optionally signed real number with a decimal point, rounded to d digits after the decimal point. If the number is less than one, the processor may place a zero in front of the decimal point. At least one zero must be output if no other digits would appear. If the number does not fit into the output field, the entire field is filled with asterisks.

Example:

```
read (unit=12, fmt="(f10.2, f10.3)") x, y
```

If the input field is

```
bbbb6.42181234567890
```

the values assigned to `x` and `y` are 6.4218 and 1234567.89, respectively. The value of d (indicating two digits after the decimal point) is ignored for `x` because the input field contains a decimal point.

Example:

```
real :: price = 473.25
print "(a, f0.2)", "The price is $", price
```

produces the output

```
The price is $473.25
```

The es Edit Descriptor

`es` is the exponential form scientific edit descriptor.

Input: The form is the same as for `f` editing.

Output: The output of the number is in the form of scientific notation; 1 ≤ |mantissa| < 10, except when the output value is 0.

Example:

```
write (unit=*, fmt="(es12.3)") b
```

The form of the output field is

[±] $y.x_1x_2 \ldots x_d$ *exp*

where y is a nonzero digit and *exp* is a signed integer; the sign must be present in the exponent.

Examples of output using the es edit descriptor are found in Table 11-4.

Table 11-4 Examples of output using the es edit descriptor

Internal value	Output field using ss, es12.3
6.421	6.421E+00
–0.5	–5.000E-01
0.0217	2.170E-02
4721.3	4.721E+03

The en Edit Descriptor

en is the engineering form scientific edit descriptor.

Input: The form is the same as for f editing.

Output: The output of the number is in the form of engineering notation; 1 ≤ |mantissa| < 1000, except when the output value is 0. The exponent is a multiple of 3.

Example:

```
write (unit=output_unit, fmt="(en12.3)") b
```

The form of the output field is

[±] $y.x_1x_2 \ldots x_d$ *exp*

where y is a one-, two-, or three-digit integer and *exp* is a signed integer that is a multiple of 3; the sign must be present in the exponent.

Examples of output using the en edit descriptor are found in Table 11-5.

Table 11-5 Examples of output using the en edit descriptor

Internal value	Output field using ss, en12.3
6.421	6.421E+00

Table 11-5 *(Continued)* Examples of output using the `en` edit descriptor

Internal value	Output field using `ss, en12.3`
–0.5	–500.000E-03
0.0217	21.70E-03
4721.3	4.721E+03

Complex Editing

Editing of complex numbers requires two real edit descriptors, one for the real part and one for the imaginary part. Different edit descriptors may be used for the two parts. Data read for a complex quantity is converted by the rules of conversion for assignment to complex. Other controls and characters may be inserted between the specification for the real and imaginary parts.

Example:

```
complex, dimension(2) :: cm
read (unit=*, fmt="(4es7.2)") cm(:)
write (unit=*, fmt="(2 (f7.2, a, f7.2, a))")  &
   real(cm(1)), " + ", aimag(cm(1)), "i ",  &
   real(cm(2)), " + ", aimag(cm(2)), "i "
```

If the input record is

```
bb55511bbb2146bbbb100bbbb621
```

the values assigned to `cm(1)` and `cm(2)` are 555.11+21.46*i* and 1+6.21*i*, respectively, and the output record is

```
b555.11b+bbb21.46ibbbb1.00b+bbbb6.21ib
```

Logical Editing

The edit descriptor used for logical editing is

```
lw
```

w is the field width.

Input: The input field for a logical value consists of any number of blanks, followed by an optional period, followed by `t` or `f`, either uppercase or lowercase, followed by anything. Valid input fields for true include `t`, `True`, `.TRUE.`, `.T`, and `thursday_afternoon`, although the last is poor practice.

Output: The output field consists of *w* – 1 leading blanks, followed by `T` or `F`.

Example:

```
write (unit=a_unit, fmt="(2l7)") l1, l2
```

If `l1` and `l2` are true and false, respectively, the output record will be

*bbbbbb*T*bbbbbb*F

Character Editing

The edit descriptor for character editing is

a [*w*]

w is the field width measured in characters. If *w* is omitted, the length of the data object being read in or written out is used as the field width. Let *len* be the length of the data object being read or written.

Input: If *w* is greater than *len*, the rightmost *len* characters in the input field are read. If *w* is less than *len*, the input is padded with blanks on the right.

Output: If *w* is greater than *len*, blanks are added on the left. If *w* is less than *len*, the leftmost *w* characters will appear in the output field. Unlike numeric fields, asterisks are not written if the data does not fit in the specified field width.

Example:

```
character(len=*), parameter :: slogan="Save the river"
write (unit=*, fmt="(a)") slogan
```

produces the output record

Save*b*the*b*river

Derived-Type Editing

Derived-type editing is described in 9.5.

Position Editing

Position edit descriptors control tabbing left or right in the record before the next list item is processed. The edit descriptors for tabbing are:

t*n*	tab to position *n*
tl*n*	tab left *n* positions
tr*n*	tab right *n* positions

n must be an unsigned integer constant with no kind parameter.

The t*n* edit descriptor positions the record just before character *n*, so that if a character is put into or taken from the record, it will be the *n*th character in the record. tr*n* moves right *n* characters. tl*n* moves left *n* characters.

If, because of execution of a nonadvancing input/output statement, the file is positioned within a record at the beginning of an input/output statement, left tabbing may not position that record any farther left than the position of the file at the start of the input/output operation.

Input: The t descriptor may position either forward or backward. A position to the left of the current position allows input to be processed twice.

Output: The positioning does not transmit characters and does not by itself cause the record to be shorter or longer. Positions that are skipped are blank filled, unless filled later in the processing. A character may be replaced by the action of subsequent descriptors, but the positioning descriptors do not carry out the replacement.

Examples: if x = 12.66 and y = –8654.123,

```
print "(f9.2, tr6, f9.3)", x, y
```

produces the record

```
bbbb12.66bbbbbb-8654.123
```

```
print "(f9.2, t7, f9.3)", x, y
```

produces the record

```
bbbb12-8654.123
```

Slash Editing

The current record is ended when a slash is encountered in a format specification. The slash edit descriptor consists of the single slash character (/).

Input: If the file is connected for sequential access, the file is positioned at the beginning of the next record. The effect is to skip the remainder of the current record. For direct access, the record number is increased by one. A record may be skipped entirely on input.

Output: If the file is connected for sequential access, the file is positioned at the beginning of a new record. For direct access, the record number is increased by one, and this record becomes the current record. An empty record is blank filled.

Example: if a = 1.1, b = 2.2, and c = 3.3,

```
print "(f5.1, /, 2f6.1)", a, b, c
```

produces two records:

```
bb1.1
bbb2.2bbb3.3
```

Colon Editing

The colon edit descriptor consists of the single colon character (:).

If the list of items in the formatted read or write statement is exhausted, a colon stops format processing at that point. It has no effect if there is more data.

Example:

```
fmt_spec = "(3(f3.1, :, /))"
write (unit=*, fmt=fmt_spec) a, b, c
```

produces only three records

```
1.1
2.2
3.3
```

The slash edit descriptor causes only three records to be output because the output list is exhausted when the colon edit descriptor is processed the third time. Without the colon edit descriptor, the output above would be followed by a blank line.

Sign Editing

Sign editing applies to numeric fields only; it controls the printing of the plus sign. It only applies to output. The sign edit descriptors are

`s`	optional plus is processor dependent
`sp`	optional plus must be printed
`ss`	optional plus must not be printed

The `s` edit descriptor indicates that the printing of an optional plus sign is up to the processor; it is the default. `sp` indicates that an optional plus sign must be printed. `ss` indicates that an optional plus sign must not be printed. The occurrence of these descriptors applies until another one (`s`, `sp`, `ss`) is encountered in the format specification.

Example: if `x(1)` = 1.46 and `x(2)` = 234.1217,

```
write (unit=*, fmt="(sp, 2f10.2)") x(1:2)
```

produces the record

bbbbb`+1.46`*bbb*`+234.12`

Decimal Symbol Editing

The edit descriptor `dp` indicates that a period is to be used for a decimal point in numeric input/output; `dc` indicates that a comma should be used. `dp` is the default.

```
print "(dc, f8.2)", 6666.22
```

produces the output

b`6666,22`

Exercise

1. Print out the message returned by `iomsg` when attempting to read 1.23 (that is a period between the 1 and 2) with an `f4.2` edit descriptor and the `decimal` option set to `comma`, either by using the `decimal` specifier or the `dc` edit descriptor.

List-Directed Formatting

List-directed formatting, also called **default formatting**, is selected by using an asterisk (*) in place of an explicit format specification in a `read`, `write`, or `print` statement. List-directed editing occurs based on the type of each list item.

Example:

```
read (unit=input_unit, fmt=*) a, b, c
```

Some rules and restrictions relating to list-directed formatting are:

1. List-directed formatting cannot be used with direct access or nonadvancing input/output.
2. The record consists of values and value separators.
3. If there are no list items, an input record is skipped or an output record that is empty is written.

Values: The values allowed are

null	a null value as in ,, (no value between separators)
c	an unsigned literal constant without a kind parameter
*r***c*	*r* repetitions of the constant *c*
*r**	*r* repetitions of the null value

where *r* is a string of digits.

Separators: The separators allowed are

,	a comma, optionally preceded or followed by contiguous blanks
;	instead of a comma, if the edit descriptor `DC` is in effect
/	a slash, optionally preceded or followed by contiguous blanks
	one or more blanks between two nonblank values

Input: Input values generally are accepted as list-directed input if they are accepted in explicit formatting with an edit descriptor. There are some exceptions. They are

1. The type must agree with the next item in the list.
2. Embedded blanks are not allowed, except within a character constant and around the comma or parentheses of a complex constant.
3. Complex items in the list include the parentheses for a complex constant. They are not treated as two reals, as is done with data edit descriptors. Blanks may occur before or after the comma. An example is

   ```
   (1.2, 5.666)
   ```

4. Logical items must not use value separators as the optional characters following the `t` or `f`.

5. Character constants must be delimited by quotes. When a character constant is continued beyond the current record, the end of the record must not be between any quotes that are doubled to indicate a quote in the character string. Value separators may be representable characters in the constant.

6. If *len* is the length of the next input list item, *w* is the length of a character constant in the input, and if

   $len \le w$ the leftmost *len* characters of the constant are used
   $len > w$ the *w* characters of the constant are used and
   the field is blank filled on the right

   Null values: A null value is encountered if

1. there is no value between separators

2. the record begins with a value separator

3. the *r** form is used.

Rules and restrictions:

1. An end of record does not signify a null value.

2. The null value does not change the next list item; however, the following value will be matched with the following list item.

3. In complex number input, the entire constant may be null, but not one of the parts.

4. If a slash terminates the input, the rest of the list items are treated as though a null value had been read. This applies to the remaining items in an array.

   Example:

```
real x(4)
read (unit=5, fmt=*) i, x(:)
```

If the input record is

```
b6,,2.418 /
```

the result is that `i` = 6, `x(1)` is unchanged, and `x(2)` = 2.418. `x(3)` and `x(4)` are unchanged.

Output: List-directed output uses the same conventions that are used for list-directed input. There are a few exceptions that are noted below for each of the intrinsic types. Blanks and commas are used as separators except for certain character constants that may contain a separator as part of the constant. The processor begins new records as needed, at any point in the list of output items. A new record does not begin in the middle of a number, except that complex numbers may be separated between the real and the imaginary parts. Very long character constants are the exception; they may be split across record boundaries. Slashes and null values are never output. Each new record begins with a blank for printing control, except for continued delimited character constants. The processor has the option of using the repeat factor, $r*c$.

Integer: The effect is as though a suitable `i`w edit descriptor were used.

Real: The effect is as though an `f` or an `es` edit descriptor were used, except that for `es` form output the first significant digit is just right of the decimal point instead of to its left. The output result depends on the magnitude of the number, and the processor has some discretion in this case.

Complex: The real and imaginary parts are enclosed in parentheses and separated by a comma (with optional blanks surrounding the comma). If the length of the complex number is longer than a record, the processor may separate the real and imaginary parts on two records.

Logical: List-directed output prints `T` or `F` depending on the value of the logical data object.

Character: Character constants are output as follows:

1. Character constants are not delimited by quotes.
2. Character constants are not surrounded by value separators.
3. Only one quote is output for each quote embedded in the character constant.
4. A blank is inserted in new records for a continued character constant.

12 Object-Oriented Programming

Object-oriented programming is a programming style that, as its name implies, places emphasis on the objects or data in a program. More traditional Fortran programming tends to emphasize the process or procedures of a program, rather than the data. It takes a long time to become a true object-oriented programmer—it takes a rather different way of looking at the whole task of programming. This will not be accomplished by studying this one chapter of a book, but what we can do is cover the Fortran language features that assist in object-oriented programming and see how they can be used in a few examples.

12.1 Extended Data Types

One of the characteristics of data in many programs is that there are data types that are different yet have many properties in common. For example, if a graphics program deals with objects that are lines to be drawn on a plane (a screen), all of the objects have a position given by the coordinates of the two end points, but they might differ in that some may have a direction and others a color. In Fortran, it is possible to define a base derived type that has the properties common to a collection of data types. This type might be called `line_type`, for example. This type can be extended to other types, such as `painted_line_type` or `vector_type` and these new data types may inherit all of the properties assigned to `line_type`, but add other properties such as `r`, `g`, and `b` to specify the red, green, and blue components or the color for a line with color and `direction` for a vector.

The following code illustrates how these derived types are defined. First the base type `line_type` is defined. All the derived types have a name that ends with `type` in order to serve as a reminder that these are names of types, not variables. The types have the `public` attribute because it is assumed that they will be in a module and be accessible by programs that use the module.

```
type, public :: line_type
   real :: x1, y1, x2, y2
end type line_type

type, public, extends(line_type) :: painted_line_type
   integer :: r, g, b   ! Values each 0-100
end type painted_line_type
```

```
type, public, extends(line_type) :: vector_type
   integer :: direction  ! 0 not directed, 1 toward (x1, y1) or 2
end type vector_type
```

An object vv of type vector_type inherits the components of the type line_type and so has five components: x1, y1, x2, y2, and direction. They are referenced as vv%x1, vv%y1, vv%x2, vv%y2, and vv%direction.

The three derived types constitute a **class**; the name of the class is the name of the base type line_type.

The effect of defining these extensible types could be achieved without extensibility as follows. First, define the type line_type as above. Then define the vector_type as:

```
type, public :: vector_type
   type(line_type) :: line
   integer :: direction
end type vector_type
```

and similarly for the painted_line type.

Now, if vv is declared to be type vector_type, it still has five components, but they are referenced as vv%line%x1, vv%line%y1, etc., and vv%direction. Although it is a minor inconvenience to have to refer to the components with a slightly longer name, the feature of extensible types would not be in the language if that were its only advantage. The real power of extensible types becomes apparent when dealing with a variable that can assume values of different data types (either a painted_line or a vector, for example). Such variables are called polymorphic.

Exercises

1. The type person_type was defined in 6.2. The components of this type are suitable for recording personal information about your friends or colleagues. Extend this type to a type employee_type by adding components suitable for recording information in a business environment. Possibilities are phone extension, mail stop, and employee identification number. Do not add a component for salary.

2. Extend the employee_type in Exercise 1 to two data types, salaried_worker_type and hourly_worker_type. In the former, include the component weekly_salary and in the latter include the components hourly_wage, overtime_factor, and hours_worked.

12.2 Polymorphism

A real variable in a Fortran program may assume different *values* at different times when the program is executing. It is also possible that a variable assumes different *types* at different times. Such a variable is called **polymorphic** (poly—many, morph—form). For example, it may be desirable to have a variable `line` that is at one time a painted line and, later in the program, a vector. Such a variable is declared with the keyword `class` instead of `type`. A polymorphic object is dynamic by nature and so must be declared to have either the `allocatable` or `pointer` attribute.

```
class(line_type), allocatable :: line
```

This variable may be assigned a value that is a plain line, a painted line, or a vector. For example

```
line = vector_type(1.1, 2.2, 4.4, 5.5, 2)
```

Note: In Fortran 2003 it was a rule that a polymorphic variable may not appear on the left side of an assignment statement; this rule was still enforced by most compilers at the time of publication of this book. However, since polymorphic variables are dynamic, they also may be created and assigned a value with the `allocate` statement. For example

```
allocate (line, source=vector_type(1.1, 2.2, 4.4, 5.5, 2))
   ! a vector from (1.1., 2.2) to (4.4, 5.5)
```

Suppose we now want to define a type of line with both color and direction. It might seem that we could create a type that extends both `painted_line_type` and `vector_type`, but this multiple inheritance is not allowed in Fortran. The best solution is probably to extend the vector type to one that includes both the direction and the color. Since a direction equal to 0 indicates no direction, this type includes all the information needed and the `painted_line_type` can be ignored.

```
type, public, extends(vector_type) :: fancy_line_type
   integer :: r, g, b
end type fancy_line_type
```

If the variable `fancy_line` is declared by

```
class(line_type), allocatable :: fancy_line
```

it has all of the line components discussed so far—end points, color, and direction. It can be set to represent an undirected blue line along the x-axis with one end at the origin by the statement

```
fancy_line = fancy_line_type(0.0, 0.0, 0.0, 1.1, 0, 0, 0, 100)
```

12.3 Procedures and Derived Types

In keeping with the idea of focusing more on the data in a program, it is possible to associate a procedure with a type. This can be done in two ways: by having a procedure (pointer) as a component of a derived type or by having a contains section within a type definition followed by a procedure.

Type-Bound Procedures

Continuing with our line examples, `length` provides an example of a type-bound procedure. Every line in the class `line_type` has a length. It is convenient to compute the length of any member of the class `line_type` with the same function, regardless of the type of the line. It is possible to achieve this with a generic function, but using a type-bound procedure means writing just one function, instead of one for each data type.

First, `line_type` is enhanced by adding the procedure `length`.

```
type, public :: line_type
   real :: x1, y1, x2, y2
contains
   procedure, public :: length
end type line_type
```

Then the function `length` is defined within the module that also contains the type definition for `line_type`. It could also be in a submodule as in 8.1.

```
function length(ab) result(length_result)

   class(line_type), intent(in) :: ab
   length_result = sqrt( (ab%x1-ab%x2)**2 + (ab%y1-ab%y2)**2 )

end function length
```

Because this procedure occurs within the type definition for `line_type`, it has a special property. If `x` is any variable of class `line_type`, it can be called by the name `x%length`, a notation just like that for selecting a component of `x`.

```
class(line_type) :: x
x = vector_type(5.3, 2.7, 6.8, 4.3, 1))
print *, x%length()
```

Exercises

1. Given an argument that is an array of objects of type `employee_type`, write a type-bound procedure that prints the name and identification number of each employee in the array.

2. Put the definition of the derived types `person_type`, `employee_type`, `hourly_worker_type`, and `salaried_worker_type` in a module and add the procedure written in Exercise 1 to the module. Write a main program to test it.

Procedure Pointer Components

A procedure pointer (10.1) may be a component of a derived type. This allows the procedure component of different objects of the type to be associated with different procedures, whereas a type-bound procedure is fixed for the type.

```
type, public :: t
   real :: x
   procedure (f), pointer, nopass :: fp
end type t
   . . .
type (t) :: tp
tp%x = 9.9
tp%fp => g   ! g must be a function with the same interface as f
   . . .
print *, tp%fp(tp%x)
```

The pass/nopass Attributes

Both procedure components and type-bound procedures have a special property: Unless otherwise indicated by using the `nopass` attribute (see the example immediately above), whenever the procedure is called, the structure which contains the procedure is passed as the first argument, without being explicitly written. See the `length` example above, in which the function `length` is called passing `x` as its argument by writing `x%length()`. The default `pass` attribute is used by the type-bound procedures of the traffic simulation in the next section. Thus, in the `print` statement in the example above, an argument must be provided to the function `tp%fp` because it has the `nopass` attribute.

12.4 Case Study: Traffic Queues

Most of what has been shown so far in this chapter could be accomplished almost as conveniently without extended types and polymorphism. The heart of this example is a queue of objects that can individually be of different data types.

Traffic Flow Simulation

Suppose the task at hand is to simulate the flow of traffic on a rectangular grid of streets. In object-oriented programming, we want to think of the types of data objects needed first, then determine what operations or procedures are to be performed. Some possibilities are vehicles, intersection, traffic lanes, pedestrians, and traffic lights. We will concentrate on vehicles and queues of vehicles at an intersection.

Vehicles

There are several sorts of vehicles involved in the traffic simulation, so a base type `vehicle_type` is needed. This type can be extended to the various types of vehicles in the class of vehicles.

To define the vehicle class, we need to decide what characteristics of every vehicle need to be recorded in a variable of the class. This will depend on the requirements of the simulation, but suppose we need at least its weight, number of wheels, and license plate number. With this much decided, the type `vehicle_type` might be defined as follows. The type definitions and all the procedures in this section will go in a module, so access should be specified explicitly for everything.

```
module vehicle_module

   implicit none
   private

   type, public :: vehicle_type
      real :: weight
      integer :: number_of_wheels
      character(len=9) :: license
   end type vehicle_type
      . . .
```

The next task is to decide which properties of each of the types of vehicles needs to be recorded. To keep the example fairly simple, we include the carrying capacity of a truck, a logical component to record whether or not a car is a taxi, and the number of passengers for a bus. All these type definitions extend the type `vehicle_type` and will complete the module `vehicle_module`.

```
      . . .
   type, public, extends(vehicle_type) :: car_type
      logical :: is_a_taxi
   end type car_type

   type, public, extends(vehicle_type) :: truck_type
      real :: capacity
   end type truck_type

   type, public, extends(vehicle_type) :: bus_type
      integer :: passengers
   end type bus_type

end module vehicle_module
```

Vehicle Queues

A **queue** is list of objects much like a linked list (10.3), except that the first object placed in the queue is the first one removed and each object is removed or processed in the order in which it was placed in the queue.

One of the parts of simulating traffic on a grid will consist of keeping track of the vehicles that are queued up at each intersection. The grid will have many intersections and we assume a rectangular grid of two-lane streets, so that each intersection has four queues. For this example, we will look at just a single queue that will be one building block of the whole program.

We are going to construct a queue of vehicles—some cars, some trucks, and some buses. Thus, an object in the queue can be of any of these three types within the class of vehicles. Without polymorphism, all objects in the queue would have to be the same type.

The development of the code for a traffic queue could occur in two phases. First, the general type for a queue could be defined, then a type for a traffic queue could extend that type. To keep the example reasonable, we will combine these two phases and construct directly a queue type that represents queues of vehicles.

With a stack, elements are inserted and removed from the same end of the list, so only one pointer to the top of the stack is needed (10.3). With a queue, an element is inserted at one end and deleted at the other end, so if pointers were used to implement the queue, two pointers would be needed—one to the head of the queue and one to its tail. To see how things can be done differently, in this example an array is used to represent the queue; the array is an array of vehicles, which are represented by a polymorphic type. In this example the elements of the array are structures containing one component, which is the derived type. This extra level of indirection is not needed if polymorphic assignment is available.

The code to process traffic queues is placed in a separate module. First we need type definitions for a vehicle queue and for each node of the queue. The procedures to process the queue are bound to the type `q_type`.

```
module v_q_module

   use vehicle_module
   implicit none
   private

   type :: node_type
      class(vehicle_type), allocatable :: v
   end type node_type

   type, public :: q_type
      private
      type(node_type), dimension(:), allocatable :: vehicles
   contains
      procedure :: empty, is_empty, insert, remove, print_licenses
   end type q_type
```

The next step is to write the procedures that process queues.

To create an empty queue, simply set the array of vehicles to an empty array. Note that q is declared `class(q_type)`, not `type(q_type)`. This is because it is a passed argument of the subroutine and a passed argument must be polymorphic. This is also true of the other procedures in the module.

```
subroutine empty(q)
   class(q_type), intent(out) :: q
   q%vehicles = [ node_type :: ]
end subroutine empty
```

The function that tests whether or not a vehicle queue is empty is simple—it tests if the array of vehicles is size 0.

```
function is_empty(q) result(is_empty_result)
   class(q_type), intent(in) :: q
   logical :: is_empty_result
   is_empty_result = (size(q%vehicles) == 0)
end function is_empty
```

The subroutine to insert a vehicle at the end of a queue is almost as simple. Simply append the vehicle as the last element of the array of vehicles.

```
subroutine insert(q, dv)
   class(q_type), intent(in out) :: q
   class(vehicle_type), intent(in), allocatable :: dv
   q%vehicles = [ q%vehicles, node_type(dv) ]
end subroutine insert
```

The **remove** subroutine sets its argument `first` to the first vehicle in the queue (if there is one) and it also has an intent **out** argument that indicates whether or not a vehicle was found in the queue, that is, whether or not the queue was empty. To remove the first element of the array, the array is set to the section consisting of elements 2 to the end of the array.

```
subroutine remove(q, first, found)
   class(q_type), intent(in out) :: q
   class(vehicle_type), allocatable, intent(out) :: first
   logical, intent(out) :: found
   found = .not. is_empty(q)
   if (.not. found) return  ! Q is empty
   first = q%vehicles(1)%v
   q%vehicles = q%vehicles(2:)
end subroutine remove
```

The subroutine `print_licenses` uses a **do** loop to print the license number of each vehicle.

```
subroutine print_licenses(q)
   class(q_type), intent(in) :: q
   integer :: n
```

```
      do n = 1, size(q%vehicles)
         select type (temp_v=>q%vehicles(n)%v)
            type is (car_type)
               write (unit=*, fmt="(a9)", advance="no") "Car:"
            type is (bus_type)
               write (unit=*, fmt="(a9)", advance="no") "Bus:"
            type is (truck_type)
               write (unit=*, fmt="(a9)", advance="no") "Truck:"
            class default
               write (unit=*, fmt="(a9)", advance="no") "Vehicle:"
         end select
         print *, trim(q%vehicles(n)%v%license)
      end do
   end subroutine print_licenses

   end module v_q_module
```

Note the use of the `select type` construct; as its name implies, a block of code is selected for execution based on the type of `q%vehicles(n)%v`. The select type construct also would be needed if any component of an extended type that is not a component of the base type were to be accessed (printed, for example).

Some of the output in this example uses a `write` statement because the long form is needed when nonadvancing input/output (11.2) (`advance=no`) is used. The asterisk in the `print` statement (*) designates list-directed or default format, whereas the format in the `write` statements is explicit.

This same task could be performed using the procedure `remove` to get the first vehicle in the queue, then printing the appropriate component and continuing until the queue is empty. This, of course, destroys the queue, which may not be desirable.

What we have built can be tested with a main program that performs various operations consisting of inserting, removing, testing, and printing. Polymorphic assignment was not available for testing, so the `allocate` statement was used instead.

```
   program test_q
      use vehicle_module
      use v_q_module
      implicit none
      class(vehicle_type), allocatable :: v
      type(q_type) :: q
      logical :: f

      call q%empty()
      print *, "Is Q empty?", q%is_empty()
      allocate (v, source=car_type(2000.0, 4, "C-1455", .false.))
      print *, "Inserting car C-1455"
      call q%insert(v)
      deallocate (v)
      print *, "Is Q empty?", q%is_empty()
      print *, "Printing Q:"
      call q%print_licenses()
```

```
      print *
      allocate (v, source=bus_type(9000.0, 6, "B-6700", 70))
      print *, "Inserting bus B-6700"
      call q%insert(v)
      deallocate (v)
      allocate (v, source=truck_type(9000.0, 18, "T-8800", 20000.00))
      print *, "Inserting truck T-8800"
      call q%insert(v)
      deallocate (v)
      allocate (v, source=bus_type(8000.0, 6, "B-6701", 70))
      print *, "Inserting bus B-6701"
      call q%insert(v)
      deallocate (v)
      print *, "Printing Q:"
      call q%print_licenses()
      print *
      print *, "Removing first vehicle in Q:"
      call q%remove(v, f)
      print *, "Found:", f, trim(v%license)
      print *, "Printing Q:"
      call q%print_licenses()
      print *
      print *, "Removing all vehicles from Q:"
      call q%empty()
      print *, "Printing Q:"
      call q%print_licenses()
      call q%remove(v, f)
      print *,f

   end program test_q
```

The output from this program follows.

```
 Is Q empty? T
 Inserting car C-1455
 Is Q empty? F
 Printing Q:
     Car: C-1455

 Inserting bus B-6700
 Inserting truck T-8800
 Inserting bus B-6701
 Printing Q:
     Car: C-1455
     Bus: B-6700
   Truck: T-8800
     Bus: B-6701

 Removing first vehicle in Q:
 Found: T C-1455
```

```
Printing Q:
    Bus: B-6700
  Truck: T-8800
    Bus: B-6701

Removing all vehicles from Q:
Printing Q:
F
```

We have made a good start toward building some of the objects and procedures needed for a traffic flow simulation, but we have a long way to go. Other modules with other data types and procedures can be built upon this work. However, the routines we have created so far illustrate some of the important concepts of object-oriented programming: type extension (inheritance), polymorphism (a queue of objects of different types), and type-bound procedures.

Exercises

1. Create a data type called `traffic_light_type` with component `color` and a subroutine `change_to` that changes the color of the light to the argument of the subroutine.

2. Define a data type `intersection` with components consisting of four traffic queues and two traffic lights (one for north–south traffic and one for east–west traffic).

3. Write a procedure that will change the direction of the green light at an intersection.

4. Write a procedure that takes an intersection as its argument. For the two traffic queues facing a green light, remove a random number of from one to four vehicles from the queues. For the two traffic queues facing a red light, insert a random number from one to three vehicles into the queues.

5. Use the procedures in Exercises 3 and 4 to simulate traffic at one intersection for 100 steps, each consisting of changing the direction of the light. For one of the traffic queues, compute the average number of vehicles in the queue over the 100 steps. After each 10 steps, print the contents of the queue being averaged to see if the results are reasonable.

6. Modify the traffic queue program in this section to remove one level of indirection by defining the type `q_type` to be

```
type, public :: q_type
   private
   type(vehicle_type), dimension(:), allocatable :: vehicles
contains
   procedure :: empty, is_empty, insert, remove, print_licenses
end type q_type
```

The type `node_type` is not needed, but modifying the subroutines in the module requires using polymorphic assignment.

7. Write a program to print the length of each of an array of objects of class `line_type` using the declarations and procedures of 12.1. Use the `select_type` construct to print the type of the line along with its length.

13 Coarrays

Many modern computers have more than one processing unit. Even an inexpensive laptop usually has several "cores". Large computer clusters may have many thousands of them. The total elapsed running time of a program often can be shortened if different parts of the program can run simultaneously on different processing units.

The use of coarrays allows programs to run faster by spreading the computation to two or more processors. This type of computation is called *Single Program Multiple Data* (SPMD).

The purpose of this chapter is to show how to use Fortran coarrays by introducing some of the basic features.

13.1 Images

In coarray Fortran, each process is called an **image**. The images are numbered 1, 2, ..., n.

An important thing to keep in mind is that each image executes the same program. Because the program code can depend on which image is executing it, the images may be executing different parts of the program at the same time or executing the same part of a program at different times.

Each image has its own copy of the program and the data. The program is the same on each image. The *execution* of the program may be different on each image.

The number of images might be set at compile, load, or execute time, depending on the system being used. This is typically controlled by a compiler option or an environment variable.

The intrinsic function `num_images` returns the number of images.

The intrinsic function `this_image` returns the number of the image on which the code is executing.

13.2 A Simple Program Using Coarrays

Here is a simple example that doesn't do anything useful, except show how coarrays work. Its features are explained in the following subsections.

```
program hello

   implicit none

   print *, "Hello from", this_image(), &
            "out of", num_images(), "images."

end program hello
```

The output is

```
Hello from           3 out of          4 images.
Hello from           1 out of          4 images.
Hello from           2 out of          4 images.
Hello from           4 out of          4 images.
```

Varying the Execution on Images

Because the intrinsic function `this_image` indicates which image the code is executing on, `select case` or `if` constructs may be used to vary the execution on different images.

```
program trig

   implicit none
   real :: x = 0.5
   character(len=*), parameter :: &
         fmt = "(a, f0.1, a, f0.5)"

   x = 0.1 + this_image()/10.0

   select case (this_image())
      case (1)
         print fmt, "sine(", x, ") = ", sin(x)
      case (2)
         print fmt, "cosine(", x, ") = ", cos(x)
      case (3)
         print fmt, "tangent(", x, ") = ", tan(x)
   end select

end program trig
```

The output is

```
tangent(.4) = .42279
sine(.2) = .19867
cosine(.3) = .95534
```

13.3 Coarray Declarations

The trigonometry example illustrates how the different images can execute different parts of a program and with different data values, but in most realistic cases the data on one image needs to be accessible on a different image. Declaring something to be a coarray allows its values on one image to be referenced on another image.

```
real, dimension(100), codimension[*] :: ca, cb, cd
real :: dimension(0:9, 4:12), codimension[0:*] :: c2
integer, codimension[*] :: n
```

The first of these declarations indicates that `ca`, `cb`, and `cd` are each arrays with 100 real elements. A copy of each of the arrays will be stored on each image. The `codimension` attribute indicates that the values of each array on each image may be accessed by the other images.

The second statement declares `c2` to be a two-dimensional array and a one-dimensional coarray with lower bound 0. The upper bound in the last codimension for each nonallocatable coarray must be *.

The last declaration indicates that `n` is a **coarray scalar**. It is not an array, but its values on each image my be accessed by the other images. For a nonallocatable scalar, the codimension is always *.

All dimensions and codimensions for an allocatable array must be `:`.

The number of dimensions plus the number of codimensions must be less than or equal to 15.

13.4 Referencing a Value on Another Image

The bracket notation used like an array subscript refers to a value on another image. Without the bracket, a value refers to the value on the local image.

The presence of the brackets indicates data movement from one image to another and may indicate a performance penalty. For example, `c2(:,:)[3]` refers to the entire array `c2` on image 3 and the values of `c2` will be copied to the current image.

In the following program, the value of `n` on image 1 is accessed by image 2 and stored as the value of `n` on image 2.

```
program ref_image_value

   implicit none

   integer, codimension[*] :: n = -99

   n = this_image()
```

```
   if (this_image() == 2) then
      print *, "Before assignment, n =", n
      n = n[1]
      print *, "After assignment, n =", n
   end if

end program ref_image_value
```

The output is

```
 Before assignment, n =            2
 After assignment, n =            1
```

13.5 The sync all Statement

The **sync all statement** causes execution on each image to wait at that point in the computation until all of the images reach that point.

Look again at the previous example, remembering that the program is being executed by all images in whatever order they get to the instructions. It is possible, but in this case not likely, that image 2 will reach the assignment statement

```
n = n[1]
```

before n as been given the value of its image on image 1. This is fixed by putting the statement

```
sync all
```

just before the if statement. Then image 2 will not execute the if construct and access n[1] until it has been set on image 1.

Other examples of the sync all statement occur later in this chapter.

13.6 Input and Output

Each image has its own units and file connections. The default output unit (*) is defined for all images. The default input unit (*) is defined only for image 1. The output from images is merged in a processor-defined manner; this means that even if output from the images is controlled to execute in order, the output itself may not appear in a file in any particular order.

13.7 Verifying Speedup

The purpose of this section is just to verify that code can be executed on different images at the same time. To verify this, we do four sorts, first one on each image, then all four sequentially on one image. The sorting program is an interchange sort, which is very inefficient. But the efficiency of the sorting algorithm is not important; what is important is that the same sort is performed each time so that comparing the times makes sense.

The system_clock Intrinsic Subroutine

The intrinsic subroutine `cpu_time` was used in 8.2 to time a matrix multiplication program. This subroutine is not appropriate for timing a program running on multiple processors because it records the total time for all processors, whereas it is the elapsed time that is significant. The intrinsic subroutine `system_clock` records the system time elapsed since the beginning of the program.

Testing Simultaneous Execution

The sorts on four images and the four sorts on one image are timed using the `system_clock` subroutine on image 1. To time the sort using four images, the system time is recorded on image 1, then a sort is done on each image, then after a `sync all`, image 1 records the system time again and computes and prints the difference. The sorts performed on one image are done under the control of a test for image 1.

```
program sort4

   use sort_mod, only: interchange_sort
   implicit none

   integer, parameter :: N = 50000
   integer :: k
   real, dimension(N) :: a
   integer :: start, stop, counts_per_second

   if (this_image() == 1) then
      call system_clock(start, counts_per_second)
   end if

   ! Generate the same set of random numbers on all images
   call random_number(a)
   call interchange_sort(a)
   print *, this_image(), a(1), a(N), all(a(:N-1)<=a(2:))
   sync all

   if (this_image() == 1) then
      call system_clock(stop)
```

```
      print *, "For 4-images, time = ", &
            (stop - start) / counts_per_second, " seconds"
   end if

   if (this_image() == 1) then
      call system_clock(start)
      do k = 1, num_images()
         call random_number(a)
         call interchange_sort(a)
      end do
      call system_clock(stop)
      print *
      print *, "For 4 sorts on 1 image,  time = ", &
            (stop - start) / counts_per_second, " seconds"
      print *, a(1), a(N), all(a(:N-1)<=a(2:))
   end if

end program sort4
```

The following output shows the times recorded when running the program on an i5-4440 four-processor system. The printout includes the first and last element of the array after sorting to verify that the smallest number is close to 0 and the largest is close to 1. The logical value indicates that the numbers are indeed sorted, as checked by the `all` intrinsic function in the `print` statement. Note that exactly the same numbers are sorted on each image, with the same results. The sorting on image 1 does take about four times as long as using four images.

```
           3  1.1205208E-05  0.9998947      T
           2  1.1205208E-05  0.9998947      T
           4  1.1205208E-05  0.9998947      T
           1  1.1205208E-05  0.9998947      T
For 4-images, time =              3  seconds

For 4 sorts on 1 image,  time =            13  seconds
 6.6114590E-06  0.9999950      T
```

13.8 A More Realistic Sorting Example

What we probably want to do in a more realistic situation is sort one large array by dividing up the work between several images. To accomplish this, the first half of the array is sorted on image 1, the second half is sorted on image 2, and the two halves of the array are merged. As before, the interchange sort (not shown) is used.

Here is the function that merges two sorted arrays. At each point in the merging process, the first elements of each sorted list are compared and the smaller one is selected for inclusion as the next element in the merged list. The process is made a little more complicated by handling the merge after one of the lists is exhausted.

```
module sort_mod
   ! subroutine interchange_sort(a)

   function merge2(a, b) result(m)

      real, dimension(:), intent(in) :: a, b
      real, dimension(size(a)+size(b)) :: m
      integer :: ka, kb, km

      ka = 1; kb = 1; km = 1

      do
         if (ka > size(a)) then
            m(km:) = b(kb:)
            return
         else if (kb > size(b)) then
            m(km:) = a(ka:)
            return
         else if (a(ka) < b(kb)) then
            m(km) = a(ka)
            km = km + 1; ka = ka + 1
         else
            m(km) = b(kb)
            km = km + 1; kb = kb + 1
         end if
      end do

   end function merge2

end module sort_mod
```

In the program sort2, random numbers are generated, stored in the array a, and copied into the array b. To do the sort that uses two images, the latter half of a is copied to image 2; then the first half is sorted on image 1 and the second half is sorted at the same time on image 2; then the sorted latter half is copied back from image 2 to image 1; then the two halves of the array are merged on image 1.

To compare doing the sort on one image, the same steps are followed, but using just one image. The two halves are sorted and then merged into one array. This process must be used because simply using the interchange sort on image 1 to sort the whole array takes longer and would not be a fair comparison of the methods.

```
program sort2

   use sort_mod
   implicit none

   integer, parameter :: N = 200000
   real, dimension(N), codimension[*] :: a
   real, dimension(N) :: b
```

```
   integer :: start, stop, counts_per_second

   if (this_image() == 1) then
      call random_seed()
      call random_number(a)
      b = a
      call system_clock(start, counts_per_second)
   end if
   sync all

   if (this_image() == 2) then
      a(N/2+1:) = a(N/2+1:)[1]
   end if
   sync all

   select case (this_image())
      case (1)
         call interchange_sort(a(:N/2))
      case (2)
         call interchange_sort(a(N/2+1:))
   end select
   sync all

   if (this_image() == 1) then
      a(N/2+1:) = a(N/2+1:)[2]
      a = merge2(a(:N/2), a(N/2+1:))
      call system_clock(stop)
      print *, "For 2-image sort,  time = ", &
            (stop - start) / counts_per_second, " seconds"
      print *, a(1), a(N), all(a(:N-1)<=a(2:))
   end if

   if (this_image() == 1) then
      call system_clock(start)
      call interchange_sort(b(:N/2))
      call interchange_sort(b(N/2+1:))
      b = merge2(b(:N/2), b(N/2+1:))
      call system_clock(stop)
      print *
      print *, "For 1-image sort,  time = ", &
            (stop - start) / counts_per_second, " seconds"
      print *, b(1), b(N), all(b(:N-1)<=b(2:))
   end if

end program sort2
```

Here is the result of running this program. Note that the sorting time using two images is one-half of the time using one image.

```
For 2-image sort,  time =    14  seconds
 5.8775768E-06  0.9999985     T

For 1-image sort,  time =    28  seconds
 5.8775768E-06  0.9999985     T
```

The performance monitor shown in Figure 13-1 provides more proof that execution is taking place on two of the computer's processors. It is shown while executing the program `sort2`, which was compiled specifying two images. Note that CPUs 1 and 4 are running at 100%, while the other two are almost idle.

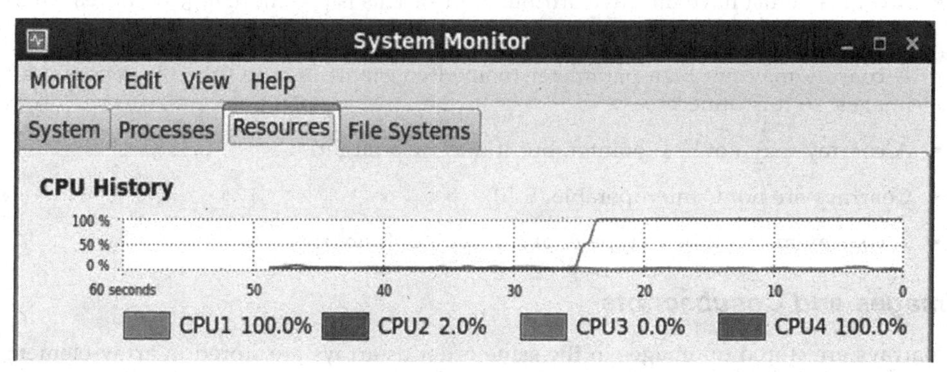

Figure 13-1 Performance monitor showing two of four processors busy.

13.9 Compiling a Program with Coarrays

Some compilers may require a special option to compile a program with coarrays. For example, with Gfortran, it is `-fcoarray`; with Intel ifort it is `-coarray` (Linux) or `/Qcoarray` (Windows).

13.10 Definition of Coarray

A **coarray** is an entity declared with the codimension attribute. It is not a coarray if it has

- a cosubscript
- a vector subscript
- an allocatable component selection.

- a pointer component selection

A **coindexed object** has a cosubscript (brackets).

These definitions are important because the rules are different for coarrays and coindexed objects.

Some Rules and Restrictions

- A coarray may be initialized.
- A coarray must have the `save` attribute (3.13). This is one more important reason to put declarations in a module!
- A coarray may not be a parameter (named constant). It would be the same on all images, so it is unnecessary.
- A coarray may not be a pointer, but it may be a target.
- Coarrays are not C interoperable (8.10).
- A coarray may not be of type `c_ptr` or type `c_funptr`.

Images and Cosubscripts

Coarrays are stored on images in the same order as arrays are stored in array-element ordering. Assume, for example, that there are five images with the declaration

```
real, codimension [2, *] :: A
```

Then the correspondence between cosubscripts and images is as follows:

`A[1, 1]` is stored on image 1
`A[2, 1]` is stored on image 2
`A[1, 2]` is stored on image 3
`A[2, 2]` is stored on image 4
`A[1, 3]` is stored on image 5

Coindexed Object Rules

A **coindex** (square brackets) indicates that the value of the object on another image is accessed.

- There are no "co-triplets"; `x[1:n]` is invalid.
- The cosubscripts must indicate a valid image. For example, if `A` is declared

  ```
  real, codimension[2,*] :: A
  ```

  and the number of images is 5, then `A[1,3]` on image 5 is valid, but `A[2,3]` is invalid because it would be on image 6, which does not exist.

Cobounds and Cosubscripts Functions

The following little program illustrates how images and cosubscripts are related. It uses the intrinsic functions lcobound and ucobound.

```
program image_fns

   implicit none
   real, codimension[0:1, *] :: C

   if (this_image() == 2) then
      print *, "num_images =", num_images()
      print *, "lower cobounds of C", lcobound(C)
      print *, "upper cobounds of C", ucobound(C)
      print *, "cosubscripts of C on image 2", this_image(C)
   end if

end program image_fns
```

Running the program produces the following.

```
num_images =           4
lower cobounds of C            0            1
upper cobounds of C            1            2
cosubscripts of C on image 2            1            1
```

Because there are four images and C is declared with codimensions [0:1, *], C is represented on the four images as follows:

C[0, 1] is on image 1
C[1, 1] is on image 2, hence the cosubscripts of C on image 2 are 1 and 1
C[0, 2] is on image 3
C[1, 2] is on image 4

The cosize of a coarray is the number of images.

13.11 Synchronization

The following cause synchronization of execution of programs on some selection of images. Some of these statements are discussed later and used in examples.

- sync statements
- lock and unlock statements
- critical and end critical statements (not discussed in this book)
- stop and end program statements

- allocation and deallocation of a coarray.

The sync all statement was discussed earlier (13.5).

Note that allocatable arrays are deallocated implicitly in several circumstances and so synchronization may take place at that time.

Syncing Images

The sync images statement causes a select set of images to suspend execution until other images have reached that point in the execution of the program.

The following program forces four images to execute a piece of code in order of image number by executing a sync between each image and the image whose number is one higher.

This program uses a coarray scalar k, which is initialized to 1 on all images, but the value of k is changed only on image 1. There is also a coarray p, but like k, its values are changed only on image 1.

To understand how this program works, it is helpful to think first about the execution by image 1, then image 2, and so forth. For image 1, the first if statement is false, so the two assignment statements are executed next. They set p(1) to 1 on image 1 and increment k on image 1 to 2. Then it issues a sync with image 2, so image 1 waits at this point until image 2 syncs with it.

Meanwhile, image 2 executes the statement that syncs it with image 1, so it waits until image 1 has executed the assignment statements described above and syncs with image 2. Then image 1 can continue to the sync all statement and wait for the other images to reach that point. Image 2 executes the two assignment statements, setting p(2) to 2 and k to 3 (both on image 1 only).

The execution on images 3 and 4 is similar. After all images reach the sync all statement, image 1 prints the four values of the array p, which are 1, 2, 3, and 4.

```
program order_images

   implicit none
   integer :: me, n_i
   integer, codimension[*] :: k = 1
   integer, dimension(4), codimension[*] :: p

   me = this_image()
   n_i = num_images()

   if (me > 1) sync images (me - 1)
      p(k[1])[1] = me
      k[1] = k[1] + 1
   if (me < n_i) sync images (me + 1)

   sync all
   if (this_image() == 1) print *, p

end program order_images
```

Syncing with All Other Images

The statement `sync images(*)` allows one image to synchronize with all of the other images. Since image 1 often executes code that the other images do not (such as reading input data using unit *), this allows such action to complete before any of the other images execute code that depends on it. The following shows how images 2 and 3 can be made to wait until some task performed by image 1 is completed. Note that the `sync images(1)` statement is needed for all images other than image 1 so that they will sync with image 1 as required by image 1's `sync images(*)` statement.

```
program sync_star

   implicit none

   integer, codimension[*] :: cointeger = 99

   select case (this_image())
   case (1)
      cointeger = 10
      sync images (*)
   case (2, 3)
      sync images (1)
      print *, this_image() * cointeger[1]
   case default
      ! Image 1 hangs without this:
      sync images (1)
      print *, cointeger
   end select

end program sync_star
```

13.12 Allocatable Coarrays

A coarray is declared allocatable by using the keyword `allocatable` and by making all dimensions and codimensions a colon (`:`).

The `allocate` statement specifies all bounds and cobounds. The last upper cobound in the `allocate` statement must be (*). The allocation on all images must be the same.

The allocation (and deallocation) causes a synchronization. Everything waits until the (de)allocation has completed on all images.

No coarray intrinsic assignment is allowed that requires a reallocation.

13.13 Case Study: Heat Transfer II

A plate has a constant heat source applied to its boundary. What is the steady state heat condition within the plate? A program to solve this problem illustrated a simple use of the associate construct in 4.6. This program is a little more general, using parameters to specify the size of the plate and allocatable arrays. But the main difference is that it times the execution of the program and compares it with the execution time for a program that uses coarrays on four images.

First, various parts of the program are examined. Later the entire program is presented.

The code to solve this on one image is a little different from the previous version in that the boundary conditions are not stored in the array representing the plate itself, but instead are stored in elements surrounding the plate (Figure 13-2). That is, the plate is considered to have two extra rows and two extra columns. A parameter `P` represents the size of the plate (also `Q = P/2` will be used in the coarray version).

```
integer, public, parameter :: P = 1000, Q = P/2
```

1.0	1.0	0.9	0.8	0.7	0.6	0.5	0.4	0.3	0.2	0.1	0.0
1.0											0.0
1.0											0.0
1.0											...
1.0											
1.0											
1.0											
1.0											
1.0											
1.0											
1.0											
1.0	0.0	0.0	...								

Figure 13-2 A 10-by-10 plate with a border all around

The part of the program that runs on one image is similar to the example in 4.6, so only the four-image code is discussed here.

In order to speed up the computation, the plate is divided into quadrants. The iterations needed to solve the heat transfer problem are carried out on each quadrant simultaneously on different images.

Although the computations for each quadrant can be executed independently of the other quadrants, some parts of the border of each quadrant are cells in an adjacent quadrant. This is illustrated by looking at the lower-left quadrant (the southwest quadrant). The northern border of this quadrant consists of cells in the northwest quadrant and the eastern border consists of cells in the southeast quadrant. Thus, the values along the southern border of the northwest quadrant must be copied to the image processing the southwest quadrant. This is the shaded area in Figure 13-3.

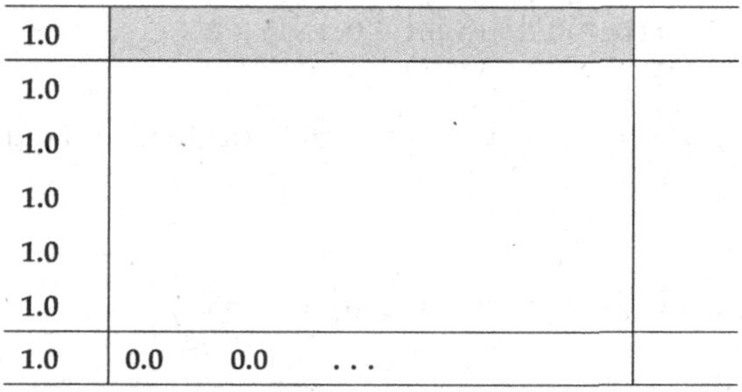

1.0			
1.0			
1.0			
1.0			
1.0			
1.0			
1.0	0.0	0.0	. . .

Figure 13-3 The southwest (SW) quadrant of the plate

Similar declarations are provided for the four-image case, except that there is a coarray **quad** representing each of four quadrants of the plate and there is a coarray scalar **diff** that keeps track of how the process is converging. Most of this code is in a module.

```
module heat_xfer_mod

   implicit none
   private

   integer, public, parameter :: P = 1000, Q = P/2
   real, public, parameter :: tolerance = 1.0e-5
   real, public, dimension(:, :), allocatable :: plate
   real, public, dimension(:, :), codimension[:,:], &
         allocatable, target :: quad
   real, public, dimension(:, :), allocatable :: temp_interior
   real, public, codimension[*] :: diff
   enum, bind(C)
      enumerator :: NW=1, SW, NE, SE
   end enum
   real, public, pointer, dimension(:,:) :: n, e, s, w, interior
   real, public, allocatable, dimension(:) :: top, bottom, left, right
   integer :: j, image
   integer, public :: n_iter = 0, alloc_stat
   public :: set_boundary_conditions, initialize_quadrants, heat_xfer
```

```
   public :: print_plate
```

The module procedure `set_boundary_conditions` allocates the arrays `top`, `left`, `right`, and `bottom` and gives them values. A different way to write this code would be to make these parameters.

```
contains

subroutine set_boundary_conditions ()

  allocate (top (0:P+1), bottom(0:P+1), &
            left(0:P+1), right (0:P+1), &
            stat = alloc_stat)
  if (alloc_stat > 0) then
     print *, "Allocation of boundary failed on image", &
              this_image()
     stop
  end if

  top = [ 1.0, ( real(j)/P, j = P, 0, -1) ]
  left = 1.0
  right = 0.0
  bottom = 0.0

end subroutine set_boundary_conditions
```

Another procedure in the module allocates the arrays for each quadrant of the plate. Remember that this same code will be executed on each image. `Q` is half of `P`.

```
subroutine initialize_quadrants ()

   allocate (quad(0:Q+1, 0:Q+1) [2,*], &
             stat = alloc_stat)
   if (alloc_stat > 0) then
      print *, "Allocation of quadrant failed on image", &
               this_image()
      stop
   end if

   allocate (temp_interior(1:Q, 1:Q), &
             stat = alloc_stat)
   if (alloc_stat > 0) then
      print *, "Allocation of temp interior failed on image", &
               this_image()
      stop
   end if
```

Next, the boundary values are set for each quadrant. Note that NW, SW, NE, and SE are simply parameters with values 1, 2, 3, and 4, declared using enumerators (1.2). The parameter names help to understand the code a little bit better.

```
! Set up boundaries
quad = 0.0
select case (this_image())
   case(NW)
      quad(:,0) = left(:Q+1)
      quad(0,:) = top(:Q+1)
   case(SW)
      quad(:,0) = left(Q:)
      quad(Q+1,:) = bottom(:Q+1)
   case(NE)
      quad(Q+1,:) = right(:Q+1)
      quad(0,:) = top(Q:)
   case(SE)
      quad(Q+1,:) = right(Q:)
      quad(Q+1,:) = bottom(Q:)
end select
```

To use the southwest quadrant as the example again, the statement

```
quad(:,0) = left(Q:)
```

assigns the lower half of the values on the left edge of the plate to the values on the left edge of the southwest quadrant and the statement

```
quad(Q+1,:) = bottom(:Q+1)
```

assigns values to the bottom of the SW quadrant from the bottom boundary of the plate.

The heat transfer computation itself consists of updating the quadrant boundaries and averaging the temperature at each point in the interior. The loop repeats until there is convergence. This code must use cosubscripts, rather than image numbers.

```
subroutine heat_xfer()

   heat_xfer_loop: do

      ! Update interior quadrant boundaries
      ! Plate boundaries have not changed
      select case (this_image())
         case(NW)
            quad(Q+1, 1:Q) = quad(1,   1:Q) [2,1] ! S
            quad(1:Q, Q+1) = quad(1:Q, 1  ) [1,2] ! E
         case(SW)
            quad(0,   1:Q) = quad(Q  , 1:Q) [1,1] ! N
            quad(1:Q, Q+1) = quad(1:Q, 1  ) [2,2] ! E
         case(NE)
            quad(Q+1, 1:Q) = quad(1,   1:Q) [2,2] ! S
            quad(1:Q, 0  ) = quad(1:Q, Q  ) [1,1] ! W
         case(SE)
            quad(0,   1:Q) = quad(Q  , 1:Q) [1,2] ! N
            quad(1:Q, 0  ) = quad(1:Q, Q  ) [2,1] ! W
```

```
      end select
```

For the southwest quadrant, values are copied from the image to the north into its top boundary, and values are copied from the image to the east into its eastern boundary. The associate construct is used when updating the cells in order to make the code a little more readable.

```
      sync all

      associate ( &
            interior => quad(1:Q, 1:Q), &
            n => quad(0:Q-1, 1:Q  ), &
            s => quad(2:Q+1, 1:Q  ), &
            e => quad(1:Q   , 2:Q+1), &
            w => quad(1:Q   , 0:Q-1))

         temp_interior = (n + e + s + w) / 4

         diff = maxval(abs(interior - temp_interior))
         interior = temp_interior
      end associate

      sync all
```

Synchronization statements are used to make sure the updating and computing are done in the correct order. Then image 1 checks the maximum of the iteration differences on the four images and exits the loop if there is convergence.

```
         if (this_image() == 1) then
            n_iter = n_iter + 1
            do image = 2, num_images()
               diff = max (diff, diff[image])
            end do
         end if
         sync all
         if (diff[1] < tolerance) exit heat_xfer_loop

      end do heat_xfer_loop

   end subroutine heat_xfer

   end module heat_xfer_mod
```

The program heat4 uses the procedures in the module to set up the computation, calls the heat transfer calculation, and sets the final computed values from each quadrant in the array plate.

```
   program heat4

      use heat_xfer_mod
```

```
   implicit none
   . . .

   call set_boundary_conditions()
   sync all
   call system_clock(start, counts_per_second)
   call initialize_quadrants()
   sync all
   call heat_xfer()
   sync all
   . . .

   if (this_image() == 1) then
      . . .
      plate(0:Q,   0:Q)   = quad(0:Q,   0:Q  ) [1,1] ! NW
      plate(Q+1:, 0:Q)   = quad(1:Q+1, 0:Q  ) [2,1] ! SW
      plate(0:Q,   Q+1:) = quad(0:Q,   1:Q+1) [1,2] ! NE
      plate(Q+1:, Q+1:) = quad(1:Q+1, 1:Q+1) [2,2] ! SE
      . . .
   end if

end program heat4
```

This program was run on a Cray XC40 using four images.

```
For 4-image solution,  time =  6  seconds

Number of iterations (4 images): 32836

For 1-image solution,  time =  37  seconds

Number of iterations (1 image): 32836

Max difference between methods: 0.
```

Changing the parameter `P` to 10 and printing the plate yields

```
1.00 1.00 0.90 0.80 0.70 0.60 0.50 0.40 0.30 0.20 0.10 0.00
1.00 0.95 0.86 0.77 0.67 0.57 0.48 0.38 0.28 0.19 0.09 0.00
1.00 0.92 0.83 0.74 0.64 0.55 0.45 0.36 0.27 0.18 0.09 0.00
1.00 0.90 0.81 0.71 0.61 0.52 0.43 0.34 0.25 0.17 0.08 0.00
1.00 0.89 0.78 0.68 0.58 0.49 0.40 0.31 0.23 0.15 0.08 0.00
1.00 0.88 0.76 0.65 0.54 0.45 0.36 0.29 0.21 0.14 0.07 0.00
1.00 0.86 0.72 0.60 0.50 0.41 0.33 0.25 0.18 0.12 0.06 0.00
1.00 0.83 0.68 0.55 0.44 0.35 0.28 0.21 0.16 0.10 0.05 0.00
1.00 0.78 0.60 0.47 0.37 0.29 0.22 0.17 0.12 0.08 0.04 0.00
1.00 0.69 0.49 0.35 0.27 0.20 0.16 0.12 0.08 0.05 0.03 0.00
1.00 0.50 0.29 0.20 0.14 0.11 0.08 0.06 0.04 0.03 0.01 0.00
0.00 0.00 0.00 0.00 0.00 0.00 0.00 0.00 0.00 0.00 0.00 0.00
```

Here is the complete program.

```
module heat_xfer_mod

   implicit none
   private

   integer, public, parameter :: P = 1000, Q = P/2
   real, public, parameter :: tolerance = 1.0e-5
   real, public, dimension(:, :), allocatable :: plate
   real, public, dimension(:, :), codimension[:,:], &
         allocatable, target :: quad
   real, public, dimension(:, :), allocatable :: temp_interior
   real, public, codimension[*] :: diff
   enum, bind(C)
      enumerator :: NW=1, SW, NE, SE
   end enum
   real, public, allocatable, dimension(:) :: top, bottom, left, right
   integer :: j, image
   integer, public :: n_iter = 0, alloc_stat
   public :: set_boundary_conditions, initialize_quadrants, heat_xfer
   public :: print_plate
   integer, public, parameter :: chunk = 100

contains

subroutine set_boundary_conditions ()

  allocate (top (0:P+1), bottom(0:P+1), &
            left(0:P+1), right (0:P+1), &
            stat = alloc_stat)
  if (alloc_stat > 0) then
     print *, "Allocation of boundary failed on image", this_image()
     stop
  end if

  top = [ 1.0, ( real(j)/P, j = P, 0, -1) ]
  left = 1.0
  right = 0.0
  bottom = 0.0

end subroutine set_boundary_conditions

subroutine initialize_quadrants ()

   allocate (quad(0:Q+1, 0:Q+1) [2,*], &
             stat = alloc_stat)
   if (alloc_stat > 0) then
      print *, "Allocation of quadrant failed on image", &
             this_image()
      stop
   end if
```

```
   allocate (temp_interior(1:Q, 1:Q), &
            stat = alloc_stat)
   if (alloc_stat > 0) then
      print *, "Allocation of temp interior failed on image", &
               this_image()
      stop
   end if

   quad = 0.0
   select case (this_image())
      case(NW)
         quad(:,0) = left(:Q+1)
         quad(0,:) = top(:Q+1)
      case(SW)
         quad(:,0) = left(Q:)
         quad(Q+1,:) = bottom(:Q+1)
      case(NE)
         quad(Q+1,:) = right(:Q+1)
         quad(0,:) = top(Q:)
      case(SE)
         quad(Q+1,:) = right(Q:)
         quad(Q+1,:) = bottom(Q:)
   end select

end subroutine initialize_quadrants

subroutine heat_xfer ()

   integer :: k

   heat_xfer_loop: do

      ! Update interior quadrant boundaries
      ! Plate boundaries have not changed
      select case (this_image())
      case(NW)
         quad(Q+1, 1:Q) = quad(1,   1:Q) [2,1] ! S
         quad(1:Q, Q+1) = quad(1:Q, 1  ) [1,2] ! E
      case(SW)
         quad(0,   1:Q) = quad(Q  , 1:Q) [1,1] ! N
         quad(1:Q, Q+1) = quad(1:Q, 1  ) [2,2] ! E
      case(NE)
         quad(Q+1, 1:Q) = quad(1,   1:Q) [2,2] ! S
         quad(1:Q, 0  ) = quad(1:Q, Q  ) [1,1] ! W
      case(SE)
         quad(0,   1:Q) = quad(Q  , 1:Q) [1,2] ! N
         quad(1:Q, 0  ) = quad(1:Q, Q  ) [2,1] ! W
      end select

      sync all
```

```
      associate ( &
         interior => quad(1:Q, 1:Q), &
         n => quad(0:Q-1, 1:Q  ), &
         s => quad(2:Q+1, 1:Q  ), &
         e => quad(1:Q  , 2:Q+1), &
         w => quad(1:Q  , 0:Q-1))

      temp_interior = (n + e + s + w) / 4

      diff = maxval(abs(interior - temp_interior))
      interior = temp_interior
      end associate

      sync all

      if (this_image() == 1) then
         n_iter = n_iter + 1
         do image = 2, num_images()
            diff = max (diff, diff[image])
         end do
      end if
      sync all
      if (diff[1] < tolerance) exit heat_xfer_loop

   end do heat_xfer_loop

end subroutine heat_xfer

subroutine print_plate(x)

   real, dimension(:,:), intent(in) :: x
   integer :: line

   print *
   do line = 1, size(x, 2)
      print "(1000f5.2)", x(line, :)
   end do

end subroutine print_plate

end module heat_xfer_mod

program heat4

   use heat_xfer_mod
   implicit none

   integer :: start, stop, counts_per_second
   integer :: line
   real, dimension(:,:), allocatable, target :: a
```

```
   real, dimension(:,:), allocatable :: temp

   call set_boundary_conditions()
   sync all
   call system_clock(start, counts_per_second)
   call initialize_quadrants()
   sync all
   call heat_xfer()
   sync all

   if (this_image() == 1) then
      call system_clock(stop)
      print *, "For 4-image solution,  time = ", &
         (stop - start) / counts_per_second, " seconds"

      allocate (plate(0:P+1,0:P+1), stat = alloc_stat)
        if (alloc_stat > 0) then
           print *, "Allocation of plate failed"
           stop
        end if

      plate(0:Q,   0:Q)   = quad(0:Q,     0:Q  ) [1,1] ! NW
      plate(Q+1:,  0:Q)   = quad(1:Q+1,   0:Q  ) [2,1] ! SW
      plate(0:Q,   Q+1:)  = quad(0:Q,     1:Q+1) [1,2] ! NE
      plate(Q+1:,  Q+1:)  = quad(1:Q+1,   1:Q+1) [2,2] ! SE

      print *
      print *, "Number of iterations (4 images):", n_iter !MPI *100

!     call print_plate(plate) ! Uncomment for debugging
   end if

   if (this_image() == 1) then
      allocate (a(0:P+1,0:P+1), temp(P,P), &
                stat = alloc_stat)
        if (alloc_stat > 0) then
           print *, "Allocation of a or temp failed"
           stop
        end if

      a = 0
      a(0, :) = top
      a(:, 0) = left
      a(:, P+1) = right
      a(P+1:, 0) = bottom

      call system_clock(start)
      n_iter = 0
      associate ( &
         interior => a(1:P, 1:P), &
```

```
         n => a(0:P-1, 1:P  ), &
         s => a(2:P+1, 1:P  ), &
         w => a(1:P,   0:P-1), &
         e => a(1:P,   2:P+1))

      call system_clock(start)
      n_iter = 0
      do
         temp = (n + e + w + s) / 4
         n_iter = n_iter + 1
         diff = maxval(abs(temp - interior))
         interior = temp
         if (diff < tolerance) exit
      end do
      end associate

      call system_clock(stop)
      print *
      print *, "For 1-image solution,  time = ", &
            (stop - start) / counts_per_second, " seconds"

      diff = maxval(abs(plate(1:P, 1:P) - a(1:P, 1:P)))

      print *
      print *, "Number of iterations (1 image):", n_iter

!     call print_plate(a) ! Uncomment to see values for small plate

      print *
      print *, "Max difference between methods:", diff

   end if

end program heat4
```

13.14 Coarray Rules and Restrictions

There are many rules and restrictions involving coarrays. Some of these are described briefly in the following subsections.

Coarrays and Procedures

- An actual argument may be a coarray or a coindexed object.
- If a dummy argument is a coarray, the actual argument must be a coarray. (A coindexed object is not a coarray!)

- To invoke a procedure with a coarray dummy argument requires an explicit interface. Put the procedure in a module!
- If a dummy argument is allocatable, the actual argument must be allocatable.
- A dummy argument of an elemental procedure may not be a coarray scalar.
- A function result may not be a coarray.
- A coarray may not be an automatic array.
- A coarray may not be a `value` or `intent out` dummy argument.

Coarrays and Pointers

- A coarray may not be a pointer.
- If the target of a pointer is a coarray, both the pointer and the target must be on the same image. That is, a coarray pointer may not point to a target on a different image.

Derived-Type Coarrays

- A coarray may not have a coarray component.
- A coarray may have a component that is allocatable or a pointer.
- The pointer component of a coarray may have a different status on different images.

In the following program, the pointer component `ptr` of `s` has different targets on the two images.

```
program ptr_comp

   implicit none

   real, target :: x = 1.1, y = 2.2
   type :: s_type
      real, pointer :: ptr
   end type s_type
   type (s_type), codimension[*] :: s

   select case (this_image())
   case (1)
      s%ptr => x
      sync images (2)
   case (2)
      sync images (1)
      s%ptr => y
```

```
      print *, s[1]%ptr  ! 1.1
      print *, s%ptr     ! 2.2
   end select

end program ptr_comp
```

- A derived type coarray may have a type-bound procedure or a component that is a procedure. In the program bound_mod, the structure t is of derived type t_type, which has both a procedure component p and a type-bound procedure s2.

```
module bound_mod

   implicit none
   private

   type, public :: t_type
      procedure(s1), pointer, nopass :: p
   contains
      procedure, nopass :: s2
   end type t_type

   public :: s1, s2

contains

   subroutine s1()
     print *, 1.1
   end subroutine s1

   subroutine s2()
     print *, 2.2
   end subroutine s2

end module bound_mod

program bound

   use bound_mod
   implicit none

   type(t_type), codimension[*] :: t

   t % p => s1
   sync all

   select case (this_image())
   case (1)
      call t%p()
      call t%s2()
      call t[2]%p()
```

```
      call t[2]%s2()
   end select

end program bound
```

Polymorphic Coarrays

- A coarray may be polymorphic.
- Intrinsic assignment to a polymorphic coarray is not permitted.
- A reference to a procedure component of a polymorphic coarray on another image is not permitted.

13.15 Case Study: Job Scheduling I

In the heat transfer case study, a single task was to be performed and it could be neatly and logically divided up between several processors. In other situations, similar, but perhaps unrelated, jobs need to be run on multiple processors to achieve good performance. This is essentially what an operating system must do to schedule all the tasks assigned to it, such as compiling a Fortran program, executing a program, surfing the internet, or downloading a file.

In the following program, image 1 creates a schedule of jobs in lists on images 2, 3, and 4. Then images 2–4 simultaneously execute the jobs on their list.

The subroutine s is a procedure that is to be executed by each of the images. The tasks they will be assigned consist of executing s with a specified value of the argument. The subroutine does nothing useful, except illustrate that the task is executed.

```
module sub_mod
   use, intrinsic :: &
      iso_fortran_env, only: output_unit
   ! . . .
contains
   subroutine s(n)
      integer, intent(in) :: n
      real, dimension(n, n) :: x
      call random_number (x)
      print "(a, i3, a, i3, f6.2)", &
         "Executing job", n, " on image", this_image(), sum(x)
      flush (output_unit)
   end subroutine s
end module sub_mod
```

The module `job_list_mod` contains the definition of the derived type `job_type`, consisting of some integer data and a procedure pointer whose interface is that of `s`. It also contains a procedure `add_job` that image 1 calls to assign a job and `get_job` that allows the other images to select a job to be executed from their list. Note that `job_list` and `list_size` are coarrays (`list_size` is a coarray scalar) because image 1 must add to those lists on each image > 1.

```
module job_list_mod

   use sub_mod
   implicit none
   private

   integer, parameter :: max_list_size = 300
   type, public :: job_type
      integer :: data
      procedure(s), pointer, nopass :: proc
   end type job_type

   type (job_type), public, &
      dimension(max_list_size), &
      codimension[*] :: job_list
   integer, public, codimension[*] :: list_size

   public :: get_job, add_job

contains

   ! Get job from list on local image
   subroutine get_job(job, empty)

      type(job_type), intent(out) :: job
      logical, intent(out) :: empty

      empty = (list_size == 0)
      if (empty) return
      job = job_list(list_size)
      list_size = list_size - 1

   end subroutine get_job

   ! Put job on list_on specified image
   subroutine add_job(job, image)

      type(job_type), intent(in) :: job
      integer, intent(in) :: image

      if (list_size[image] == max_list_size) then
         print *, "Job list is full on image", image
         stop
      end if
      list_size[image] = list_size[image] + 1
      job_list(list_size[image])[image] = job
```

```
    end subroutine add_job

end module job_list_mod
```

In the program `job_list`, image 1 puts some jobs in each job list, spreading them around by using the `modulo` function to determine an image number. The `sync all` statement causes images 2 and greater to wait until image 1 has built the job lists for all the images. Then the jobs are executed on each of the images greater than 1.

```
program job_list

   use sub_mod
   use job_list_mod

      implicit none

   type (job_type) :: job
   integer :: n
   logical :: empty

   list_size = 0
   job%proc => s

   if (this_image() == 1) then
      do n = 1, 10
         job%data = n
         call add_job(job, image =  modulo(n, 4) + 1)
      end do
   end if

   sync all

   ! All images execute jobs on their job list,
   job_loop: do

      call get_job(job, empty)
      if (empty) exit job_loop
      call job%proc(job%data)

   end do job_loop

end program job_list
```

Running the program produces the following.

```
Executing job  8 on image  1 33.52
Executing job  9 on image  2 40.97
Executing job  5 on image  2 14.60
Executing job  1 on image  2  0.97
Executing job 10 on image  3 52.25
```

```
Executing job  6 on image  3 19.40
Executing job  2 on image  3  2.50
Executing job  7 on image  4 26.21
Executing job  4 on image  1  7.12
Executing job  3 on image  4  4.41
```

According to the calculation in subroutine s, the last value printed should be approximately 1/2 of the square of the job number, so things look reasonable.

13.16 Locking Statements

Locks may be set using the **lock** and **unlock statements** to restrict access to data by code on another image. For example, the statement

```
lock (list_lock[1])
```

indicates that no image except the image executing the lock statement can execute the statements that follow it until the lock is unlocked using a statement such as

```
unlock (list_lock[1])
```

The object list_lock in each of these statements is declared to be type lock_type, a derived type defined in the intrinsic module iso_fortran_env.

13.17 Case Study: Job Scheduling II

In the previous program, image 1 created job lists on each of the other images; then the other images executed all the jobs on their own list. In the following program, the job list is created and kept on image 1 and each of the other images then fetches the next task from the job list and executes it; this continues until the job list is empty.

When the get_job subroutine is executed by an image, the list lock is set so that no other image can access the list until the fetch is complete. Then it is unlocked.

The procedure to execute a job is the same as in the previous job scheduling program.

```
module job_list_mod

   use sub_mod
   use, intrinsic :: &
      iso_fortran_env, only: lock_type
   implicit none
   private
```

```
   integer, parameter :: max_list_size = 300
   type(lock_type), codimension[*] :: list_lock
   type, public :: job_type
      integer :: data
      procedure(s), pointer, nopass :: proc
   end type job_type

   type (job_type), public, &
      dimension(max_list_size), &
      codimension[*] :: job_list
   integer, public, codimension[*] :: list_size

   public :: get_job, add_job

contains
   ! Get job from list on image 1
   subroutine get_job(job, empty)

      type(job_type), intent(out) :: job
      logical, intent(out) :: empty

      lock (list_lock[1])
      empty = (list_size[1] == 0)
      if (empty) then
         unlock (list_lock[1])
         return
      end if
      job = job_list(list_size[1])[1]
      list_size[1] = list_size[1] - 1
      unlock (list_lock[1])

   end subroutine get_job

   ! Put job on list_on image 1
   subroutine add_job(job)

      type(job_type), intent(in) :: job

      if (list_size[1] == max_list_size) then
         print *, "Job list is full"
         stop
      end if
      list_size = list_size + 1
      job_list(list_size) = job

   end subroutine add_job

end module job_list_mod
```

```
program jobs

! Omitting declarations, etc.
   . . .
!  Create a list of jobs on image 1
   if (this_image() == 1) then
      do n = 1, 10
         job%data = n
!        job%data also could be different for each job
         call add_job(job)
      end do
   end if

   sync all

   ! All images > 1 execute jobs on job list,
   !     which is on image 1
   select case (this_image())
   case (2:)
   job_loop: do

      call get_job(job, empty)
      if (empty) exit job_loop
      call job%proc(job%data)

   end do job_loop
   end select

end program jobs
```

Running the program produces the following.

```
Executing job 10 on image  4 52.25
Executing job  9 on image  2 40.97
Executing job  8 on image  3 33.52
Executing job  7 on image  4 27.28
Executing job  6 on image  3 18.73
Executing job  4 on image  3  9.58
Executing job  3 on image  3  5.06
Executing job  2 on image  3  1.24
Executing job  5 on image  2 14.60
Executing job  1 on image  3  0.29
```

13.18 Exercises

1. Write a program to use two images to find the maximum value in an array of 1,000,000 random numbers. Model the program after the program `sort2` in 13.8. Compute the maximum of the first half of the array on image 1 and the maximum of the second half of the array on image 2. Then find the maximum of those two values.

   Why don't you get a significant speed improvement?

2. Write a program `sort4x` that splits an array into four parts, sorts each part on an image, then merges the four parts into the sorted array (use `merge` three times). Model it after the program `sort2` (13.8). Make sure that the one-image version does the same computation as the four-image version and compare the times.

3. Modify the heat transfer program so that the boundary values for each quadrant are updated once for every hundred computational iterations. Look in the code for the comment

   ```
   ! Update interior quadrant boundaries
   ```

   and do the update and synchronization only if the value of `niter` is a multiple of 100. Does the four-image computation run faster? Why are the computational results slightly different?

4. Modify the heat transfer program so that the images form an x-by-y grid of images, rather than a 2-by-2 grid. For example, if a 16-processor system is available, it might work well to have a 4-by-4 grid of images (and it would be convenient if the dimensions of the plate were a multiple of 4). Is performance better or worse if the program uses a 2-by-8 grid of images? This last question should be easy to determine if parameters have been used properly.

5. Modify the first job scheduling program in this chapter so that image 1 sets different procedures for the other images to execute when executing the items on their job lists. In addition to the subroutine `s`, provide two or three other procedures. Note that the interface for these additional procedures must be the same as `s`.

6. Modify the second job scheduling program in this chapter so that the other images are executing the jobs while image 1 is creating them. The images executing the jobs must wait sometimes in case the job queue is empty. There must be some special signal to indicate that all the jobs have been placed on the job list.

Intrinsic Procedures A

A.1 Intrinsic Functions

An **intrinsic function** is an inquiry function, an elemental function, or a transformational function. An **inquiry function** is one whose result depends on the properties of its principal argument, rather than the value of this argument; in fact, the argument value may be undefined. An **elemental function** is one that is specified for scalar arguments but may be applied to array arguments, as described in A.2. All other intrinsic functions are **transformational functions**; they almost all have one or more array-valued arguments or an array-valued result.

The names of the intrinsic procedures are given in A.4–A.15. In most cases, they accept arguments of more than one type; for functions, the type of the result is usually the same as the type of one or more of the arguments.

Kind Arguments

Style note: Several intrinsic procedures have kind values as arguments. It is recommended that they be integer named constants (parameters); other more general forms that represent a constant for the compiler are permitted, but not recommended.

A.2 Elemental Intrinsic Procedures

Elemental Intrinsic Function Arguments and Results

For an elemental intrinsic function, the shape of the result is the same as the shape of the argument with the greatest rank. If the arguments are all scalar, the result is scalar. For those elemental intrinsic functions that have more than one argument, all arguments must be conformable (i.e., have the same shape). In the array-valued case, the values of the elements, if any, of the result are the same as would have been obtained if the scalar-valued function had been applied separately, in any order, to corresponding elements of each argument. Arguments called `kind` must always be specified as a scalar integer parameter. The value of the parameter must be a processor-supported kind number.

Elemental Intrinsic Subroutine Arguments

For an elemental intrinsic subroutine, either all actual arguments must be scalar or all `intent(out)` arguments must be arrays of the same shape, and the remaining arguments must be conformable with them. In the case that the `intent(out)` arguments are arrays, the values of the elements, if any, of the results are the same as would be obtained if the subroutine with scalar arguments were applied separately, in any order, to corresponding elements of each argument.

A.3 Positional Arguments or Argument Keywords

All intrinsic procedures may be invoked with either positional arguments or argument keywords. The descriptions in A.4–A.15 give the keyword names and positional sequence. A keyword is required for an argument only if a preceding optional argument is omitted or a preceding actual argument is specified using a keyword. For example, a reference to `cmplx` may be written in the form `cmplx (real_part, complex_part, m)` or in the form `cmplx(y=complex_part, kind=m, x=real_part)`.

Many of the argument keywords have names that are indicative of their usage. For example,

kind	Describes the kind of the result
string, string_a	An arbitrary character string
back	Indicates a string scan is to be from right to left (backward)
mask	A mask that may be applied to the arguments
dim	A selected dimension of an array argument

A.4 Argument Presence Inquiry Function

The inquiry function `present` permits an inquiry to be made about the presence of an actual argument associated with a dummy argument that has the `optional` attribute. Its result is logical.

present(a)	Argument presence

A.5 Numeric, Mathematical, Character, and Logical Procedures

Numeric Functions

The elemental functions `int`, `real`, `dble`, and `cmplx` perform type conversions. The elemental functions `aimag`, `conjg`, `aint`, `anint`, `nint`, `abs`, `dim`, `dprod`, mod, `modulo`, `floor`, `ceiling`, `sign`, `max`, and `min` perform simple numeric operations.

abs(a)	Absolute value
aimag(z)	Imaginary part of a complex number
aint(a, kind) Optional kind	Truncation to whole number
anint(a, kind) Optional kind	Nearest whole number
ceiling(a, kinf) Optional kind	Least integer greater than or equal to number
cmplx(x, y, kind) Optional y, kind	Conversion to complex type

conjg(z)	Conjugate of a complex number
dble(a)	Conversion to double precision
dim(x, y)	Difference x–y, if positive, otherwise zero
dprod(x, y)	Double precision product of two default real values
floor(a, kind)	Greatest integer less than or equal to number
Optional kind	
int(a, kind)	Conversion to integer type
Optional kind	
max(a1, a2, a3,...)	Maximum value
Optional a3,...	
min(a1, a2, a3,...)	Minimum value
Optional a3,...	
mod(a, p)	Modulo function; a–floor(a/p)*p; if nonzero, modulo(a,p) has the sign of a
modulo(a, p)	Modulo function; a–floor(a/p)*p; if nonzero, modulo(a,p) has the sign of p
nint(a, kind)	Nearest integer
Optional kind	
real(a, kind)	Conversion to real type
Optional kind	
sign(a, b)	Absolute value of a with the sign of b

Mathematical Functions

The elemental functions `sqrt`, `exp`, `log`, `log10`, `sin`, `cos`, `tan`, `asin`, `acos`, `atan`, `atan2`, `sinh`, `cosh`, `tanh` `asinh`, `acosh`, `atanh`, `gamma`, `log_gamma`, `bessel_j0`, `bessel_j1`, `bessel_jn`, `bessel_y0`, `bessel_y1`, and `bessel_yn` evaluate mathematical functions.

acos(x)	Arccosine
acosh(x)	Hyperbolic arccosine
asin(x)	Arcsine
asinh(x)	Hyperbolic arcsine
atan(x)	Arctangent
atan2(y, x)	Arctangent of y/x
atanh(x)	Hyperbolic arctangent
bessel_ j0(x)	Bessel function of the 1st kind, order 0
bessel_ j1(x)	Bessel function of the 1st kind, order 1
bessel_ jn(n, x)	Bessel function of the 1st kind, order n
bessel_ jn(n1, n2, x)	Bessel functions of the 1st kind
bessel_ y0(x)	Bessel function of the 2nd kind, order 0
bessel_y1(x)	Bessel function of the 2nd kind, order 1
bessel_yn(n, x)	Bessel function of the 2nd kind, order n
bessel_yn(n1, n2, x)	Bessel functions of the 2nd kind
cos(x)	Cosine
cosh(x)	Hyperbolic cosine
erf(x)	Error function
erfc(x)	Complementary error function
erfc_scaled(x)	Scaled complementary error function
exp(x)	Exponential
gamma(x)	Gamma function
hypot(x, y)	Euclidean distance function
log(x)	Natural logarithm
log10(x)	Common logarithm (base 10)
log_gamma(x)	Logarithm of the absolute value of the gamma function
sin(x)	Sine

sinh(x)	Hyperbolic sine
sqrt(x)	Square root
tan(x)	Tangent
tanh(x)	Hyperbolic tangent

Character Functions

The elemental functions `ichar`, `char`, `iachar`, `achar`, `index`, `verify`, `adjustl`, `adjustr`, `scan`, and `len_trim` perform character operations. The elemental functions `lge`, `lgt`, `lle`, and `llt` compare character strings based on the ASCII collating sequence. The transformational function `repeat` returns repeated concatenations of a character string argument. The transformational function `trim` returns the argument with trailing blanks removed.

achar(i, kind) Optional kind	Character in given position in the ASCII collating sequence
adjustl(string)	Adjust left; move leading blanks to end
adjustr(string)	Adjust right; move trailing blanks to beginning
char(i, kind) Optional kind	Character in given position in collating sequence
iachar(c, kind) Optional kind	Position of a character in the ASCII collating sequence
ichar(c, kind) Optional kind	Position of a character in collating sequence
index(string, substring, back, kind) Optional back, kind	Starting position of a substring
lge(string_a, string_b)	Greater than or equal based on the ASCII collating sequence
lgt(string_a, string_b)	Greater than based on the ASCII collating sequence
lle(string_a, string_b)	Less than or equal based on the ASCII collating sequence
llt(string_a, string_b)	Less than based on the ASCII collating sequence
repeat(string, ncopies)	Repeated concatenation
scan(string, set, back, kind) Optional back, kind	Scan a string for any character in a set of characters
trim(string)	Remove trailing blank characters
verify(string, set, back, kind) Optional back, kind	Find a character in a string not in a set of characters

Character Inquiry Function

The inquiry functions `len` and `len_trim` return the length of a character entity. The value of the argument to this function need not be defined. It is not necessary for a processor to evaluate the argument of this function if the value of the function can be determined otherwise. The function `new_line` returns a newline character of the same kind as its character argument `a`.

len(string, kind) Optional kind	Length of a character entity
len_trim(string, kind) Optional kind	Length without trailing blank characters
new_line(a)	A newline character the same kind as a

Logical Function

The elemental function `logical` converts between objects of type logical with different kind parameter values.

logical(l, kind) Optional kind	Convert between objects of type logical with different kind type parameters

Kind Functions

The inquiry function `kind` returns the kind parameter value of an integer, real, complex, or logical entity. The transformational function `selected_char_kind` returns the character kind parameter value of the character kind `name`; the values of name that usually are supported are `DEFAULT`, `ASCII`, and `ISO_10646`. The transformational function `selected_real_kind` returns the real kind parameter value that has at least the decimal precision and exponent range specified by its arguments. The transformational function `selected_int_kind` returns the integer kind parameter value that has at least the decimal exponent range specified by its argument.

kind(x)	Kind parameter value
selected_char_kind(name)	Kind parameter of a character kind with a given name
selected_int_kind(r)	Integer kind parameter value, sufficient for integers with r digits
selected_real_kind(p, r) Optional p, r	Real kind parameter value, given decimal precision p and range r

A.6 Numeric Manipulation and Inquiry Functions

The numeric manipulation and inquiry functions are described in terms of a model for the representation and behavior of numbers on a processor. The model has parameters that are determined so as to make the model best fit the machine on which the executable program is executed.

Models for Integer and Real Data

The model set for integer i is defined by

$$i = s \times \sum_{k=1}^{q} w_k \times r^{k-1}$$

where r is an integer exceeding 1, q is a positive integer, each w_k is a nonnegative integer less than r, and s is +1 or −1. The model set for real x is defined by

$$x = \begin{cases} 0 & \text{or} \\ s \times b^e \times \sum_{k=1}^{p} f_k \times b^{-k} \end{cases}$$

where b and p are integers exceeding 1; each f_k is a nonnegative integer less than b; f_1 is also nonzero; s is +1 or −1; and e is an integer that lies between some integer maximum e_{max} and some integer minimum e_{min} inclusively. For $x = 0$, its exponent e and digits f_k are defined to be zero. The integer parameters r and q determine the set of model integers, and the integer parameters b, p, e_{min}, and e_{max} determine the set of model floating point numbers. The parameters of the integer and real models are available for each integer and real data type implemented by the processor. The parameters characterize the set of available numbers in the definition of the model. The numeric manipulation and inquiry functions provide values related to the parameters and other constants related to them. Examples of these functions in this section use the models

$$i = s \times \sum_{k=1}^{31} w_k \times 2^{k-1}$$

and

$$x = 0 \quad \text{or} \quad s \times 2^e \times \left(1/2 + \sum_{k=2}^{24} f_k \times 2^{-k} \right) \quad -126 \le e \le 127$$

Numeric Inquiry Functions

The inquiry functions `radix`, `digits`, `minexponent`, `maxexponent`, `precision`, `range`, `huge`, `tiny`, `epsilon` and `storage_size` return scalar values related to the parameters of the model associated with the type and type parameters of the arguments. The value of the arguments to these functions need not be defined, and pointer arguments may be disassociated.

digits(x)	Number of significant digits p in the model
epsilon(x)	Number that is almost negligible compared to one
huge(x)	Largest number in the model
maxexponent(x)	Maximum exponent in the model; e_{max}
minexponent(x)	Minimum exponent in the model; e_{min}
precision(x)	Decimal precision
radix(x)	Base of the model; b for real and r for integer
range(x)	Decimal exponent range; floor(log10(huge(x)))
storage_size(a, kind) Optional kind	Storage size in bits
tiny(x)	Smallest positive number in the model

Floating Point Manipulation Functions

The elemental functions `exponent`, `scale`, `nearest`, `fraction`, `set_exponent`, `spacing`, and `rrspacing` return values related to the components of the model values associated with the actual values of the arguments.

exponent(x)	Exponent part e of a model number
fraction(x)	Fractional part of a number
nearest(x, s)	Nearest different processor number in a given direction
rrspacing(x)	Reciprocal of the relative spacing

	of model numbers near given number
scale(x, i)	Multiply a real x by its base to an integer power i
set_exponent(x, i)	Set exponent part of a number x to i
spacing(x)	Absolute spacing of model numbers near a given number

A.7 Bit Manipulation and Inquiry Procedures

The bit manipulation procedures consist of a set of several functions and one subroutine. Logical operations on bits are provided by the functions `ior`, `iand`, `not`, and `ieor`; the functions `bge`, `bgt`, `ble`, and `blt` compare bits; the functions `leadz` and `trailz` indicate the number of leading and trailing zero bits; `popcnt` and `poppar` indicate the number and parity of one bits; shift operations are provided by the functions `shifta`, `shiftl`, `shiftr`, `ishft` and `ishftc`; bit subfields may be referenced by the function `ibits` and by the subroutine `mvbits`; bit masks are constructed by `maskl` and `maskr`; `merge_bits` merges bits; single-bit processing is provided by the functions `btest`, `ibset`, and `ibclr`.

For the purposes of these procedures, a bit is defined to be a binary digit w located at position k of a nonnegative integer scalar object based on a model nonnegative integer defined by

$$j = \sum_{k=0}^{s-1} w_k \times 2^k$$

and for which w_k may have the value 0 or 1. An example of a model number compatible with the examples used in A.6 would have $s = 32$, thereby defining a 32-bit integer.

An inquiry function `bit_size` is available to determine the parameter s of the model. The value of the argument of this function need not be defined. It is not necessary for a processor to evaluate the argument of this function if the value of the function can be determined otherwise.

Effectively, this model defines an integer object to consist of s bits in sequence numbered from right to left from 0 to $s - 1$. This model is valid only in the context of the use of such an object as the argument or result of one of the bit manipulation procedures. In all other contexts, the model defined for an integer in A.6 applies. In particular, whereas the models are identical for $w_{s-1} = 0$, they do not correspond for $w_{s-1} = 1$ and the interpretation of bits in such objects is processor dependent.

bge(i, j)	Bitwise greater than or equal to
bgt(i, j)	Bitwise greater than
bit_size(i)	Number of bits in the model; s
ble(i, j)	Bitwise less than or equal to
blt(i, j)	Bitwise less than
btest(i, pos)	Bit testing
iand(i, j)	Logical and
ibclr(i, pos)	Clear bit
ibits(i, pos, len)	Bit extraction
ibset(i, pos)	Set bit
ieor(i, j)	Exclusive or

ior(i, j)	Inclusive or
ishft(i, shift)	Logical shift
ishftc(i, shift, size)	Circular shift
Optional size	
leadz(i)	Number of leading zero bits
maskl(i, kind)	Left justified mask
Optional kind	
maskr(i, kind)	Right justified mask
Optional kind	
merge_bits (i, j, mask)	Merge of bits under mask
mvbits(from, frompos, len, to, topos)	Copy bits from one object to another
not(i)	Logical complement
popcnt(i)	Number of one bits
poppar(i)	Parity of bits, 0 or 1
shifta(i, shift)	Right shift with fill
shiftl(i, shift)	Left shift
shiftr(i, shift)	Right shift
trailz(i)	Number of trailing zero bits

A.8 Array Intrinsic Functions

The array intrinsic functions perform the following operations on arrays: vector and matrix multiplication, numeric or logical computation that reduces the rank, array structure inquiry, array construction, array manipulation, and geometric location.

The Shape of Array Arguments

The transformational array intrinsic functions operate on each array argument as a whole. The shape of the corresponding actual argument must therefore be defined; that is, the actual argument must be an array section, an assumed-shape array, an explicit-shape array, a pointer that is associated with a target, an allocatable array that has been allocated, or an array-valued expression.

Some of the inquiry intrinsic functions accept array arguments for which the shape need not be defined. They include the function `lbound` and certain references to `size` and `ubound`.

Mask Arguments

Some array intrinsic functions have an optional `mask` argument that is used by the function to select the elements of one or more arguments to be operated on by the function.

The `mask` affects only the value of the function, and does not affect the evaluation, prior to invoking the function, of arguments that are array expressions.

A `mask` argument must be of type logical.

Vector and Matrix Multiplication Functions

The matrix multiplication function `matmul` operates on two matrices, or on one matrix and one vector, and returns the corresponding matrix–matrix, matrix–vector, or vector–matrix product. The arguments to `matmul` may be numeric (integer, real, or complex) or logical arrays. On logical matrices and vectors, `matmul` performs Boolean matrix multiplication (that is, multiplication is `.and.` and addition is `.or.`).

The dot product function `dot_product` operates on two vectors and returns their scalar product. The vectors are of the same type (numeric or logical) as for `matmul`. For logical vectors, `dot_product` returns the Boolean scalar product.

dot_product(vector_a, vector_b)	Dot product of two rank-one arrays
matmul(matrix_a, matrix_b)	Matrix multiplication

Array Reduction Functions

The array reduction functions `sum`, `product`, `maxval`, `minval`, `count`, `any`, `all`, `parity`, `iparity`, and `norm2` perform numerical, logical, and counting operations on arrays. They may be applied to the whole array to give a scalar result or they may be applied over a given dimension to yield a result of rank reduced by one. The optional `dim` argument selects which subscript is reduced. By use of a logical mask that is conformable with the given array, the computation may be confined to any subset of the array (for example, the positive elements).

all(mask, dim) Optional dim	True if all values are true
any(mask, dim) Optional dim	True if any value is true
count(mask, dim, kind) Optional dim, kind	Number of true elements in an array
iparity(array, dim, mask) Optional mask	Exclusive or of array elements
iparity(array, mask) Optional mask	Exclusive or of array elements
maxval(array, dim, mask) Optional dim, mask	Maximum value in an array
minval(array, dim, mask) Optional dim, mask	Minimum value in an array
norm2(x, dim) Optional dim	L_2 norm of an array
parity(mask, dim) Optional dim	True if number of elements is odd
product(array, dim, mask) Optional dim, mask	Product of array elements
sum(array, dim, mask) Optional dim, mask	Sum of array elements

Array Inquiry Functions

The functions `size`, `shape`, `lbound`, and `ubound` return, respectively, the size of the array, the shape, and the lower and upper bounds of the subscripts along each dimension. The size, shape, or bounds must be defined. The function `allocated` indicates whether an allocatable array (or scalar) is allocated. The function is_contiguous indicates whether an array is contiguous or not.

The values of the array arguments to these functions need not be defined.

allocated(array) or allocated(scalar)	Indicates whether allocatable argument is allocated
is_contiguous(array)	Test contiguity of an array

lbound(array, dim, kind) Optional dim, kind	Lower dimension bounds of an array
shape(source, kind) Optional kind	Shape of an array or scalar
size(array, dim, kind) Optional dim, kind	Total number of elements in an array
ubound(array, dim, kind) Optional dim, kind	Upper dimension bounds of an array

Array Construction Functions

The functions `merge`, `spread`, `pack`, and `unpack` construct new arrays from the elements of existing arrays. `merge` combines two conformable arrays into one array by an element-wise choice based on a logical mask. `spread` constructs an array from several copies of an actual argument (`spread` does this by adding an extra dimension, as in forming a book from copies of one page). `pack` and `unpack`, respectively, gather and scatter the elements of a one-dimensional array from and to positions in another array where the positions are specified by a logical mask.

merge(tsource, fsource, mask	Merge under mask Where mask is true, result is tsource, elsewhere result is fsource
pack(array, mask, vector) Optional vector	Pack an array into an array of rank one under a mask. Result size is count(mask) If vector is present, result is padded with terminal elements of vector to size(vector)
spread(source, dim, ncopies)	Replicates array by adding a dimension
unpack(vector, mask, field)	Unpack an array of rank one into an array under a mask. Where mask is true, elemnts of field are replaced by elements of vector; result has shape of mask

Array Reshape Function

`reshape` produces an array with the same elements as its argument, but with a different shape.

reshape(source, shape, pad, order) Optional pad, order	Reshape an array

Array Manipulation Functions

The functions `transpose`, `eoshift`, and `cshift` manipulate arrays. `transpose` performs the matrix transpose operation on a two-dimensional array. The shift functions leave the shape of an array unaltered but shift the positions of the elements parallel to a specified dimension of the array. These shifts are either circular (`cshift`), in which case elements shifted off one end reappear at the other end, or end-off (`eoshift`), in which case specified boundary elements are shifted into the vacated positions.

cshift(array, shift, dim) Optional dim	Circular shift
eoshift(array, shift, boundary, dim) Optional boundary, dim	End-off shift
transpose(matrix)	Transpose of an array of rank two

Array Location Functions

The function `findloc` finds the first location of a value. `maxloc` and `minloc` return the location (subscripts) of an element of an array that has maximum and minimum values, respectively. By use of an optional logical mask that is conformable with the given array, the reduction may be confined to any subset of the array. The size of the returned value is the rank of the array.

findloc (array, value, dim, mask, kind, back) Optional mask, kind, back	Location of a specified value
findloc (array, value, mask, kind, back) Optional mask, kind, back	Location of a specified value
maxloc(array, dim, mask, kind, back) Optional mask, kind, back	Location of maximum values in an array
maxloc(array, mask, kind, back Optional mask, kind, back	Location of maximum values in an array
minloc(array, dim, mask, kind, back) Optional mask, kind, back	Location of minimum values in an array
minloc(array, mask, kind, back Optional mask, kind, back	Location of minimum values in an array

A.9 Pointer Nullify and Association Status Inquiry Functions

The function `null` returns a null (disassociated) pointer. The function `associated` tests whether a pointer is currently associated with any target, with a particular target, or with the same target as another pointer.

associated(pointer, target) Optional target	Association status or comparison
null(mold) Optional mold	A null pointer

A.10 Type Extension Inquiry Functions

The function `extends_type_of` determines whether the dynamic type of the first argument is an extension of the dynamic type of the second argument. The function `same_type_as` determines whether the dynamic types of the two arguments are the same.

extends_type_of(a, mold)	True if the dynamic type of a is an extension of the dynamic type of mold
same_type_as(a, b)	True if the dynamic types of a and b are the same

A.11 Date and Time Subroutines

The subroutines `date_and_time` and `system_clock` return integer data from the date and real-time clock. The time returned is local, but there are facilities for finding out the difference between local time and Coordinated Universal Time.

The subroutine `cpu_time` returns in seconds the amount of CPU time used by the program from the beginning of execution of the program.

cpu_time(time)	Obtain processor time in seconds
date_and_time(date, time, zone, values) Optional date, time, zone, values	Obtain date and time date="ccyymmdd" time="hhmmss.sss" values=[year, month, day, gmt_min, hr,min,sec,msec]
system_clock(count, count_rate, count_max) Optional count, count_rate, count_max	Obtain data from the system clock count_rate is in counts per second

A.12 Pseudorandom Numbers

The subroutine `random_number` returns a pseudorandom number greater than or equal to 0.0 and less than 1.0 or an array of pseudorandom numbers. The subroutine `random_seed` initializes or restarts the pseudorandom number sequence.

random_number(harvest)	Returns pseudorandom number
random_seed(size, put, get) Optional size, put, get	Initializes or restarts the pseudorandom number generator

A.13 Transfer Procedures

The transformational function `transfer` and pure subroutine `move_alloc` transfer data without changing any bits.

move_alloc(from, to)	Transfer an allocation from one object to another of the same type
transfer(source, mold, size)	Result is the same bits as source, but interpreted with the type and type parameters of mold

A.14 Testing Input/Output Status

The elemental functions `is_iostat_end` and `is_iostat_eor` test an iostat value to determine if it indicates an end-of-file condition or an end-of-record condition, respectively.

is_iostat_end(i)	True if iostat value indicates end of file
is_iostat_eor(i)	True if iostat value indicates end of record

A.15 Command Line Manipulation

The command line manipulation procedures allow the program to inquire about the command and environment that invoked the program. `command_argument_count` is an inquiry function that gives the number of arguments in the command line. `execute_command_line`, `get_command`, `get_command_argument`, and `get_environment_variable` are subroutines.

command_argument_count()	Number of command line arguments
execute command line (command, wait, exitstat, cmdstat, cmdmsg) Optional wait, exitstat, cmdstat, cmdmsg	Execute a command line
get_command(command, length, status) Optional command, length, status	The entire command that invoked the program
get_command_argument(number, value, length, status) Optional value, length, status	The appropriate argument from the command
get_environment_variable(name, value, length, status, trim_name) Optional value, length, status, trim_name	The value of the named system environment variable

A.16 Coarray Functions

The functions `image_index`, `lcobound`, `ucobound`, `num_images`, and `this_image` are used with coarrays.

image index(coarray, sub)	Convert cosubscripts to image index
lcobound (coarray, dim, kind) Optional dim, kind	Lower cobound(s) of a coarray
num_images()	Number of images
this_image()	Image of execution
this_image(coarray, dim) Optional dim	Cosubscript(s) for this image
ucobound (coarray, dim, kind) Optional dim, kind	Upper cobound(s) of a coarray

A.17 Atomic Functions

The functions `atomic_define` and `atomic_ref` define and reference a variable atomically. These functions are not discussed in this book.

atomic define(atom, value)	Define a variable atomically
atomic ref(value, atom)	Reference a variable atomically

B Fortran Language Forms

This appendix contains an informal description of the major parts of the Fortran programming language. It is not a description of the complete Fortran language.

The notation used is a very informal variation of Backus–Naur form (BNF) in which characters from the Fortran character set are to be written as shown. Lowercase italicized letters and words represent general categories for which specific syntactic entities must be substituted in actual statements. The rules are not a complete and accurate syntax description of Fortran, but are intended to give the general form of the important language constructs.

Brackets [ ] indicate optional items. A "list" means one or more items separated by commas.

program:

```
program program name
     [ use statements ]
     implicit none
     [ declaration statements ]
     [ executable statements ]
end program program-name
```

public module:

```
module module name
     use statements
     public
end module module name
```

private module:

```
module module name
     [ use statements ]
     implicit none
     private
     [ access statements ]
     [ declaration statements ]
[ contains
     [ subroutines and functions ] ]
end module module name
```

submodule:

```
submodule ( parent ) submodule name
    [ use statements ]
    [ import statements ]
    implicit none
    private
    [ declaration statements ]
[ contains
    [ subroutines and functions ] ]
end submodule submodule name
```

subroutine:

```
[ prefix ] subroutine subroutine name ( [ argument list ] )
    [ use statements ]
    [ declaration statements ]
    [ executable constructs ]
end subroutine subroutine name
```

function:

```
[ prefix ] function function name ( [ argument list ] ) &
        result ( function result )
    [ use statements ]
    [ declaration statements ]
    [ executable constructs ]
end function function name
```

prefix:

```
elemental
impure
pure
recursive
```

use statement:

```
use module name [ , rename list ]
use module name , only : [ only list ]
```

access statement:

```
public :: list of procedures, operators, assignments
private :: list of procedures, operators, assignments
```

declaration statement:

interface block
enumerator definition

intrinsic statement
type definition statement
type declaration statement

interface block:

`interface` [*generic specification*]
  [*import statements*]
  [*interface specifications*]
`end interface`

interface specification:

procedure statement
procedure interface body

enumerator definition:

`enum, bind(c)`
  [*enumerator* :: *enumerator list*] . . .
`end enum`

enumerator:

parameter [= *scalar integer constant expression*]

intrinsic statement:

`intrinsic` :: *list of intrinsic procedure names*

type definition statement:

`type` , [*access specifier*] :: *derived type name*
  [`private`]
  component declarations
  [*contains*
    [*procedure statements*]]
`end type` *derived type name*

type declaration statement:

`type` [, *attribute list*] :: *initialization list*

initialization:

name [= *expression*]
name => *expression*

type:

`integer` [(`kind=` *kind parameter*)]
`real` [(`kind=` *kind parameter*)]
`complex` [(`kind=` *kind parameter*)]
`logical` [(`kind=` *kind parameter*)]
`character` (`len=` *length parameter*)
`character(len=*)`
`type` (*type name*)

attribute:

access specifier
`parameter`
`allocatable`
`dimension` (*array bounds*)
`intent` (*intent specifier*)
`optional`
`pointer`
`save`
`target`
`value`

access specifier:

`public`
`private`
`protected`

intent specifier:

`in`
`out`
`in out`

executable construct:

if construct
if statement
do construct
case construct
block construct
where construct
select type construct
associate construct
go to statement
continue statement
assignment statement
pointer assignment statement
allocate statement

deallocate statement
call statement
cycle statement
exit statement
statement
stop statement
sync all statement:
sync images statement:
sync memory statement:
lock statement:
unlock statement:
open statement
close statement
inquire statement
read statement
print statement
write statement
backspace statement
rewind statement
endfile statement
wait statement

if construct:

[*construct name* :] if (*logical expression*) then
    executable statements
[else if (*logical expression*) then
    executable statements]
    ...
[else
    executable statements]
end if [*construct name*]

if statement:

if (*logical expression*) *statement*

do construct:

[*construct name* :] do [*loop control*]
    executable constructs
end do [*construct name*]

loop control:

variable = *start* , *stop* [: *stride*]

case construct:

```
[ construct name : ] select case ( expression )
    case ( case selector )
            executable statements
    [ case ( case selector )
            executable statements ]
            ...
    [ case default
            executable statements ]
end select [ construct name ]
```

block construct:

```
[ construct name : ] block
    executable constructs
end block [ construct name ]
```

where construct:

```
where ( mask expression )
    where body constructs
[ elsewhere ( mask expression )
    where body constructs
    ... ]
[ elsewhere
    where body constructs ]
endwhere
```

where body construct:

assignment statement
where construct
where statement
pointer assignment statement
where construct
where statement

select type construct:

```
select type ( polymorphic variable )
    [ type guard
            executable statements ]
            ...
end select
```

type guard:

`type is (` *type* `)`
`class is (` *derived type* `)`
`class default`

associate construct:

[*construct name* :] `associate (` *association list* `)`
    [*execution constructs*]
`end associate` [*construct name*]

go to statement:

`go to` *label*

continue statement:

label `continue`

assignment statement:

variable = *expression*

pointer assignment statement:

pointer => *target*

allocate statement:

`allocate (` *allocation list* [`, stat=` *variable*] `)`

deallocate statement:

`deallocate (` *deallocation list* [`, stat=` *variable*] `)`

call statement:

`call` *subroutine name* `(` [*actual argument list*] `)`

cycle statement:

`cycle` [*do construct name*]

exit statement:

`exit` [*construct name*]

return statement:

`return`

stop statement:

`stop` [*stop code*]

sync all statement:

`sync all` [([*sync status list*])]

sync images statement:

`sync images` (*images* [, *sync status list*])

sync memory statement:

`sync memory` [([*sync status list*])]

lock statement:

`lock` (*lock variable* [, *lock status list*]

unlock statement:

`unlock` (*lock variable* [, *lock status list*]

open statement:

`open` (*open specifier list*)

close statement:

`close` (*close specifier list*)

inquire statement:

`inquire` (*inquire specifier list*)

read statement:

`read` *format* [, *variable list*]
`read` (*io control specifier list*) [*variable list*]

print statement:

`print` *format* [, *expression list*]

write statement:

`write` (*io control specifier list*) [*expression list*]

io control specifier:

`unit=` *value*
`fmt=` *value*
`rec=` *value*
`iostat=` *value*
`iomsg=` *value*
`advance=` *value*
`size=` *value*
`pos=` *value*

backspace statement:

`backspace (` *position specifier list* `)`

endfile statement:

`endfile (` *position specifier list* `)`

rewind statement:

`rewind (` *position specifier list* `)`

position specifier:

`unit=` *value*
`iostat=` *value*
`iomsg=` *value*

wait statement:

`wait (` *wait specifier list* `)`

Index

Symbols

A

B

C

D

G

H

I

J

K

L

Q

R

T